Science and Global Security Monograph Series

International Control of Tritium for Nuclear Nonproliferation and Disarmament

Science and Global Security Monograph Series

Harold Feiveson, Series Editor

Volume I
Reversing the Arms Race: How to Achieve and Verify Deep Reductions in the Nuclear Arsenals
F. von Hippel and R.Z. Sagdeev, Editors

Volume II
The Security Watershed: Russians Debating Defense and Foreign Policy after the Cold War
A.G. Arbatov, Editor

Volume III
By Fire and Ice: Dismantling Chemical Weapons while Preserving the Environment
D. Koplow

Volume IV
International Control of Tritium for Nuclear Nonproliferation and Disarmament
Martin B. Kalinowski

Science and Global Security Monograph Series

International Control of Tritium for Nuclear Nonproliferation and Disarmament

Martin B. Kalinowski

CRC PRESS

Boca Raton London New York Washington, D.C.

The views expressed herein are those of the author and do not necessarily reflect the views of the CTBTO Preparatory Commission.

Cover image: "Tritium nuclides in the cage," model by Alexander Glaser and Christoph Pistner, photograph by Martin B. Kalinowski.

Library of Congress Cataloging-in-Publication Data

Catalog record is available from the Library of Congress

Direct all inquiries to CRC Press LLC, 2000 N.W. Corporate Blvd., Boca Raton, Florida 33431.

Visit the CRC Press Web site at www.crcpress.com

International Standard Book Number 0-415-31615-4
Printed in the United States of America 1 2 3 4 5 6 7 8 9 0
Printed on acid-free paper

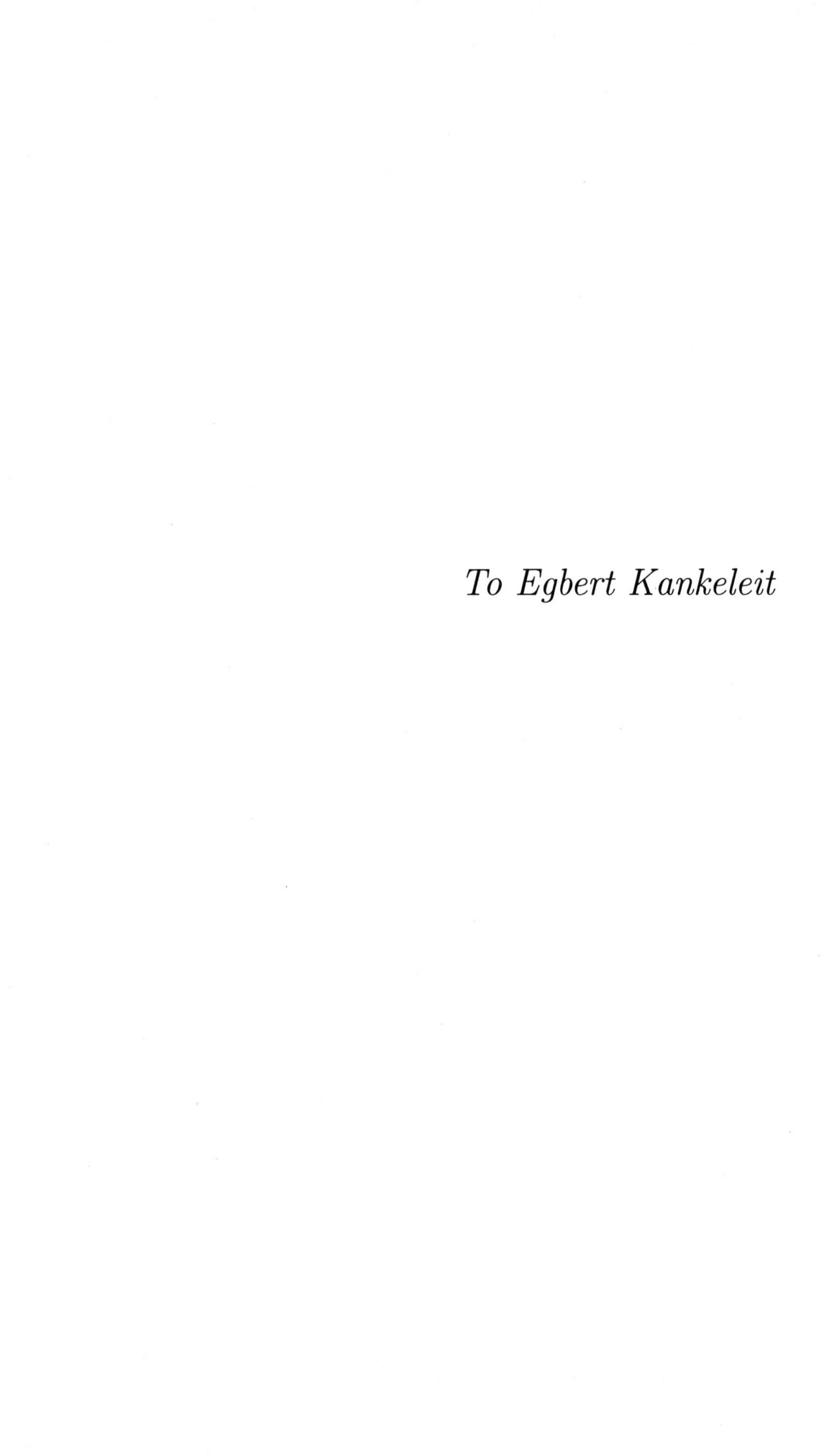

To Egbert Kankeleit

Series Preface

From the very beginning in 1989, when a small group of American and Russian scientists launched the international journal, *Science and Global Security*, we envisioned that it would be accompanied over time by a monograph series.

Like the journal, we sought in such a monograph series to publish high-quality technical and scientific analyses related to arms control and security policy. The goal was to improve the quality and cumulative impact of communication on these subjects within the international scientific community and to help create a common understanding of the technical basis for new policy initiatives. As far as possible, we aimed for the volumes to be written so that their essential conclusions can be understood by nonspecialists while containing enough technical detail so that results can be reproduced by experts.

While the journal itself has flourished and is now in its 15th year, very few of the monographs have been published to date. It is now our hope that under our new publisher, CRC Press of the Taylor and Francis Group, we can publish more monographs, some culled and thematically collected from the *Science and Global Security* archives, and some detailed and comprehensive examinations of specific topics such as the present volume's analysis of initiatives to control the production of tritium.

Harold A. Feiveson
Series Editor

Contents

Foreword

This book has converted me into a former skeptic. I am the anonymous critic whom Martin Kalinowski quotes as describing the proposal to reduce nuclear arsenals by a tritium production cutoff as an attempt to get "the tail to wag the dog." In this book, however, he has succeeded in making credible to me the idea that tritium controls could indeed provide a new dimension to nuclear.

Only a relatively small number of neutrons are needed to produce the few grams of tritium that can "boost" the explosive power of a modern nuclear warhead from a few hundred tons to a hundred thousand tons of chemical-explosive equivalent. This is only a few percent as many neutrons as would be required to produce the "significant quantity" of 8 kilograms of plutonium which the International Atomic Energy Agency has estimated would suffice for the construction of a first-generation, Nagasaki-type plutonium bomb, with a yield equivalent to about 20,000 tons of chemical explosive.

From this perspective, the challenge of verifying a cutoff of the production of tritium for weapons would be much greater than the already great challenge of verifying the Fissile Material Cutoff Treaty on which negotiations were to begin at the Conference on Disarmament in Geneva in 1999.

However, once plutonium is produced, it is, on the scale of a human lifetime, forever, unless it is destroyed by fission. Plutonium-239, the principal isotope in weapons-grade plutonium, has a half-life of 24,000 years.

By contrast, the half-life of tritium is 12.3 years. As a consequence, less than 0.3% of the tritium in the U.S. and Soviet nuclear arsenals at the end of the Cold War will still exist a century from now.

The U.S. production reactors were shut down irreversibly in 1988. By the year 2000, if we neglect other losses, the U.S. stock of tritium for weapons was only half of what it was at the end of the Cold War. Cries were heard from conservatives (and even more from political representatives of regions where a new production facility might be built) that billions of dollars should be spent to assure a source of replacement tritium. Arms controllers argued that there are still great opportunities for further reductions in the nuclear-weapon stockpiles and that any investment in new tritium-production capabilities could be delayed indefinitely. It will be five half-lives, more than 60 years, before the Cold-War tritium stockpile will have decayed to the point where it could keep at full power "only" hundreds of nuclear weapons — still enough to destroy modern civilization. In the end, it was decided to develop tritium production targets and "buy neutrons" as needed from the Tennessee Valley Authority's power-production reactors.

As Kalinowski points out, the difficulty of tritium control efforts will also depend upon the future scale of tritium's nonweapon uses. The leading industrial states are still putting more than a billion dollars per year into efforts to develop fusion reactors powered by tritium. It would take the annual fusion of about 40 million grams of tritium to produce as much fusion power as the world's fission reactors do today. At the moment, commercial fusion power is still beyond the horizon. However, its advent would represent a profound challenge to attempts to control nuclear-weapon stockpiles through tritium control—just as the fulfillment of the dream of a world fueled by plutonium would have greatly increased the difficulty of preventing the diversion of plutonium to weapons.

Tritium controls cannot serve as a substitute for fissile-material controls in efforts to achieve nuclear disarmament, but Kalinowski argues persuasively that they have a potentially complementary role. He also points out that there already exists a tritium control system between Canada and EURATOM under which all tritium supplied by Canada is subjected to verification concerning its end use.

This book provides the essential background for those who would like to explore the possibility of making such a regime global.

Frank Niels von Hippel
Princeton University

Preface

The main goal of this book is to present a qualified conclusion as to whether it is feasible from a physical and technical point of view to implement and verify an agreement on an international control of tritium for its nonuse in nuclear weapons.[1] Although precedents on a limited scale are given for international tritium control, a more comprehensive approach has been proposed as an instrument to reverse the vertical proliferation (nuclear disarmament) in recognized nuclear weapons states as well as against the horizontal proliferation of advanced nuclear weapon designs into emerging nuclear weapons states (Chapter 1).

In general, an international agreement on tritium control would have the goal to reduce — ultimately to zero — the use of tritium originating from any source, especially from civilian or undeclared military sources in nuclear weapons. The main goals of the verification of an international tritium control agreement are 1) detection of noncompliance along with confidence-building by the demonstration of compliance and 2) deterrence against noncompliance. Further goals are to increase the technical burden for potential proliferators, to clarify uncertainties, and to increase transparency of tritium production, stocks, and applications.

A comprehensive study covers all significant civilian and military uses of tritium and possible control objectives (Chapter 1). This is followed by a comprehensive diversion path analysis based on a survey of the production, occurrence, and availability of tritium in different types of nuclear facilities (Chapter 2).[2] From this, procedures to verify the nondiversion of tritium are derived and evaluated (Chapter 3).

On the basis of these findings, a **technical assessment of an international tritium control agreement** is presented (Chapter 4), which evaluates whether control is technically feasible without inappropriate costs or side-effects. Concepts and criteria developed by the International Atomic Energy Agency (IAEA) for nuclear safeguards applied to fissile materials, especially plutonium and highly enriched uranium,[3] are transferred and adapted appropriately to a concept for international tritium control. The criteria for the technical assessment are outlined in Section 3.2.1. The pivotal question is whether adequate verification can be achieved. The goals of the verification depend on the goals of the particular international agreement in question.

As a **conclusion**, this study shows that verification of international tritium control agreements both for horizontal nonproliferation of tritium and for reversing vertical proliferation would be technically feasible with an effectiveness which is politically acceptable, even though the significant quantity of tritium is defined here to be as low as 1 gram. Tritium control could be implemented without adding excessive inspection activities. Instead, existing structures and routine inspection procedures within the nuclear nonproliferation regime and others adapted from radiation protection procedures can be used.

Endnotes

1. This study was undertaken as part of an interdisciplinary project. Some special physical questions were solved by the author of this monograph in his Ph.D. thesis on *Monte Carlo Simulationen und Experimente zum zerstörungsfreien Nachweis von Lithium-6. Physikalische Fragen zur Tritiumkontrolle,* published by Shaker Verlag, Aachen, 1997. The political preconditions and implications are investigated in Colschen, L.C.: *Die Internationalisierung der Tritiumkontrolle als Baustein des Nichtweiterverbreitungsregimes für Kernwaffen. Bedingungen, Einflußfaktoren und Folgen*, Ph.D. thesis, published by Shaker Verlag, Aachen, 1998. The manuscript has been updated for publication with regard to all developments that are relevant for international tritium control as of summer 2003.

2. Most of the data presented in Chapter 2 and Appendix A are of historic nature as of 1992. They serve for illustrative purposes. No significant changes have occurred since then. All considerations regarding tritium control that are based on these historic numbers are still fully valid.

3. For brevity this is called simply "nuclear safeguards" in this monograph, and the term "control" is used for tritium instead of "safeguards."

Chapter 1

Dealing with the civilian/military ambivalence toward tritium

1.1 Introduction

Tritium has various military as well as civilian applications. The latter have always lagged behind military applications and were partly enabled or triggered by the availability of tritium from military production.

Tritium is believed to be used by all nuclear weapon states in most of their nuclear warheads. Even de facto nuclear weapons states have engaged in tritium technology. Tritium on its own is neither sufficient to produce nuclear weapons, nor a necessary component to design a simple nuclear warhead. It becomes strategically significant only when it is combined with either plutonium or highly enriched uranium-235 (HEU) in a complex design which requires a high degree of technical competence. Whereas plutonium-239 and uranium-235 are the fissile materials that provide the yield, the main military application of tritium is to fuse it with deuterium so as to release neutrons which in their turn increase ("boost") the efficiency and thus the explosive yield of a given amount of plutonium-239 and uranium-235. Due to its short half-life of 12.3 years, tritium has to be replenished on a regular basis. Since no natural resources are available, tritium has to be produced in nuclear reactors, which implies the diversion of nuclear energy for military purposes.

With the horizontal and vertical proliferation of know-how and the capacity to produce nuclear weapon, tritium and tritium technology are clearly gaining significance within the nuclear proliferation process. This is exemplified by the partially successful attempts of undeclared nuclear weapons states to acquire tritium and tritium technology or by the large investments planned by the U.S. to develop and establish a new military tritium production facility despite the progress in nuclear disarmament.

In the 1990s, the control of tritium became a topic on the political agenda which resulted in several activities to strengthen and harmonize export control measures and in a mandate for EURATOM to control tritium that is being supplied by Canada to fusion research facilities in EURATOM member states. A further prominent proposal is to consider tritium during negotiations for a verified agreement on a production cutoff for fissile materials. Tritium might even play a key role in the nuclear disarmament process due to its radioactive decay.

Neither the quantity of tritium nor its mode of production, chemical state, physical condition, or degree of purity determines or indicates an intended military or civilian use. Since it is impossible to physically differentiate between "military tritium" and "civilian tritium," the respective sociotechnological environment has to be taken into account if a judgment is required. Since physical barriers can never be completely tight, the most efficient way to prevent the diversion of tritium for military purposes, besides binding and verified political commitments, is to minimize any production and application.

1.2 Tritium and tritium technology

Tritium[1] (symbol: T or ^3H) is the superheavy isotope of hydrogen, which is the lightest element with just one proton in its nucleus. Ordinary hydrogen (H or ^1H) has no neutron; heavy hydrogen (D or ^2H), also known as deuterium, has one neutron; and tritium has two neutrons in its nucleus. Their relatively large mass differences have no significant influence on the chemical behavior of the three hydrogen isotopes. Tritium is a gas and appears in molecular form as T_2. It can substitute ordinary hydrogen in any of its compounds. Of particular importance is its oxidized form as "superheavy" water (T_2O), as tritiated water (HTO), or tritiated heavy water (DTO). The main physical data are summarized in Table 1.1.

Table 1.1 Main physical data of tritium.

atomic weight	3.017
half-life [years]	12.36
decay path	beta (100%)
decay product	^3He
maximum beta energy [keV]	18.6
average beta energy [keV]	5.7
specific activity of T_2 [TBq/g]	358
specific activity of T_2O [TBq/g]	268
specific activity of HTO [TBq/g]	53.7
volume of 1 g tritium at standard temp. and pressure [l]	3.720
decay heat [W/Bq]	9.2×10^{-16}
biological half-life [days]	8–12
dosage produced by 1 GBq in man (70 kg) [mSv/day]	1.19

The physical properties of the three hydrogen isotopes differ significantly. Their different mobilities and boiling points can be used for isotopic separation. The most remarkable characteristic of tritium as compared to the other two hydrogen isotopes is its radioactive decay. It emits a beta particle to become the stable isotope helium-3.

$$T \rightarrow {}^{3}He + \beta^{-} + \bar{\nu}_{e} + 18.6\ keV. \quad (1.1)$$

The half-life of tritium is 12.3 years, i.e., a given quantity of tritium decreases at a rate of about 5.5% per year by radioactive decay. The specific activity of pure tritium is 360 TBq/g. Due to its decay, tritium does not accumulate. In the biosphere, tritium exchanges rapidly and establishes equilibria between various systems (International Atomic Energy Agency [IAEA], 1979). It is not possible to find any water sample which does not contain tiny amounts of tritium.

The bulk of natural hydrogen is the normal hydrogen and some 0.015% is deuterium. The concentration of naturally produced tritium is one atom of tritium in 10^{18} atoms of hydrogen. Accordingly, the unit for tritium concentration has been defined as 1 TU = 10^{-18}. This corresponds to 0.120 Bq/l. The whole natural inventory of tritium in the hydrosphere amounts to a mere few kilograms.

Artificial production and emission of tritium has increased the natural concentration several times. About one hundred kilograms have been emitted from artificial sources, basically by atmospheric testing of nuclear weapons in the late 1950s and early 1960s. Table 1.2 shows the total steady state inventory of the biosphere from natural and artificial sources.

In terms of radiation safety, tritium is a major radionuclide emitted by nuclear reactors and reprocessing plants during normal operation.[2] A survey on the containment performance of various tritium facilities and the radiological impact of tritium emissions showed that the radiation dosages received by the most exposed individual during normal operation of large facilities can come close to the regulatory limits (Kalinowski, 1993). However, in the case of accidents at proposed fusion reactors which use state-of-the-art technology, compliance with regulatory limits cannot be guaranteed (Kalinowski, 1993).

The natural abundance of tritium is so low that it cannot be economically exploited. It has to be produced artificially by means of a nuclear reaction. This can be done in nuclear reactors (see Section 2.4). In thermonuclear weapons, tritium is produced in situ from lithium deuteride via the following nuclear reaction:

$${}^{6}Li + n \rightarrow \alpha + T + 4.78\ MeV. \quad (1.2)$$

Most civilian applications use the energy of the beta particle which results from the decay of tritium (see Section 1.3.1). A mixture of gaseous tritium and deuterium is used in nuclear weapons as a source of neutrons (see Section 1.3.2). For this purpose the following nuclear reaction is used:

$$D + T \rightarrow \alpha + n + 17.6\ MeV. \quad (1.3)$$

This reaction releases fusion energy, but the insertion of gaseous tritium adds little to the explosive yield of the nuclear weapon. In thermonuclear weapons, tritium is produced in situ and most of the total energy release originates from the fusion of tritium and deuterium (see Section 1.3.2).

If the amount of tritium handled exceeds a certain limit, it will be necessary to undertake a certain technological effort in order to contain the tritium and to protect workers and the environment from contamination. With the exception of some medical and tracer research applications, tritium beyond this limit is usually required.

Table 1.2 Tritium inventory in the biosphere from natural atmospheric production, nuclear weapon tests, and emissions from nuclear facilities.

source	production rate [triton/(cm^2s)] [a]	production or emission rate [g/y]	steady state content in the biosphere [kg]	remark
natural prod.	0.12–1.2	100–1000	1.8–18	review of 13 studies (1953–67) (UNSCEAR, 1977)
natural prod.	0.14–0.90	110–700	1.9–12	review of 5 studies based on cosmic ray and nuclear cross-section data (Phillips and Easterly, 1980)
natural prod.	0.25±0.08	200±64	3.5±1.1	good estimate (Craig and Lal, 1961)
natural prod.	0.14–2.0	110–1600	1.9–28	review of 5 studies based on geochemical inventory (Phillips and Easterly, 1980)
natural prod.	0.5±0.3	400±240	7.0±4.2	best estimate (Craig and Lal, 1961)
natural prod.	0.39+0.09–0.29	310+70–230	5.5+1.2–4.0	more recent study Röther (1980)
emissions	–	500–700	9–12	see footnote [b]
nuclear expl.	–	–	90–56+230	see footnote [c]
total			100–200	see footnote [d]

[a] A triton is the completely ionized nucleus of tritium. The earth surface area is 510.1×10^6 km^2.

[b] In the past, the largest contribution to tritium releases from nuclear facilities stemmed from emissions of tritiated heavy water from nuclear power plants at a rate of some 1% of the inventory per year. Reprocessing of spent fuel constitutes the second largest contribution. A significant increase in this kind of release is expected for the next ten years. See Section 2.6.9.

[c] The total release of tritium by nuclear weapons tests in the late 1950s and early 1960s was about 465 kg (United Nations Scientific Committee on the Effects of Atomic Radiation (UNSCEAR), 1977). The figure was calculated from an estimate of the total inventory in 1970 (300 kg). Allowing for decay, some 90 kg did exist at the end of 1991. The total explosive power of thermonuclear bomb tests above ground was 318 Mt TNT. Estimates for the production factor range from 190 to 1800 PBq/Mt TNT (Phillips and Easterly, 1980). Therefore, the total tritium production from thermonuclear bomb tests lies between 170 and 1600 kg, i.e., between 34 and 320 kg did still exist. Contributions from fission bomb tests are less than 4 kg.

[d] According to CFFTP (1988), the total amount in the atmosphere (0.6 kg, 0.4%), continental waters (18 kg, 12.5%), and oceans (129 kg, 90%) was 144 kg in 1986. According to UNSCEAR (1977) 7.2% can be found in the atmosphere, 27% in land surface and biosphere, and 65% in oceans.

Safe handling of tritium involves various technology components for transporting, storing, measuring, handling, and containing tritium. A minimum configuration for tritium handling encompasses storage (e.g., uranium getter storage), instruments for inventory measurements (e.g., a steel tank with pressure gauge, thermometer, and mass spectrometer), equipment for transfer processes (e.g., pipes, loading device, vacuum pump), provisions for radiation protection (glove box, emergency device to remove tritium from air), devices for emission control and monitoring (molecular sieve and catalyst for oxidizing tritium gas, stack, and room air monitor), and a waste treatment facility.

The establishment of a complete production line necessitates the use of more sophisticated tritium technology for breeding (irradiation target for nuclear reactor), purification, and isotopic separation (e.g., gas chromatograph) (see Section 2.4 and Albright and Paine, 1988).

Measurement of tritium concentrations in air for radiation monitoring is very difficult. Tritium is one of the most challenging isotopes to detect because it emits only beta rays which have an exceptionally low energy (maximum of 18.6 keV). The average travel distance of these beta particles before being absorbed is some 5 mm in air and 6 μm in water. The most widely used method to measure tritium is to apply a scintillation detector. In this case, air or water samples are mixed with a liquid scintillator. Gas proportional counters can be employed if the tritium to be measured is made to replace an ordinary hydrogen atom in the counter gas methane.[3]

1.3 The use of tritium

1.3.1 Civilian uses of tritium

Tritium is offered commercially as gas with a typical purity of about 90% (Lieser, 1980) or as tritiated water.

The development of commercial tritium applications was intensified in the early 1960s, primarily because excess amounts of tritium were made available by the U.S. Atomic Energy Commission (USAEC). USAEC made 100 g of tritium available in 1959, which was subsequently sold by the Oak Ridge National Laboratory (ORNL) for peaceful applications. In comparison, 4.1 g of tritium had been sold in 1958, and only 1.3 g in the period between 1948 and 1957 (Anonymous, 1959). Tritium was primarily welcomed as a pure beta emitter to replace radium in self-powered light sources whose gamma radiation causes unacceptable radiation doses.

The historical development of worldwide civilian demand can be seen in Figure 1.1. At the end of the 1960s, the worldwide consumption of tritium was about 20 g per year. In the early 1970s, the European tritium industry put annually about 30 g of tritium in luminous paints and 20 g in self-luminating lights ("beta-lights") (Desroches, 1973). The worldwide demand increased slightly to 100 g per year in the mid 1970s, basically for radio-luminous paints and self-luminating lights ("beta-lights"). For 1978 the amount of tritium used in consumer goods was estimated to be 100 g in Europe and 200 g worldwide (Krejci, 1979). In 1979 the commercial consumption of tritium peaked[4] at approximately 800 g, but in 1980 it dropped markedly back to about 100 g/y because safety regulations were tightened

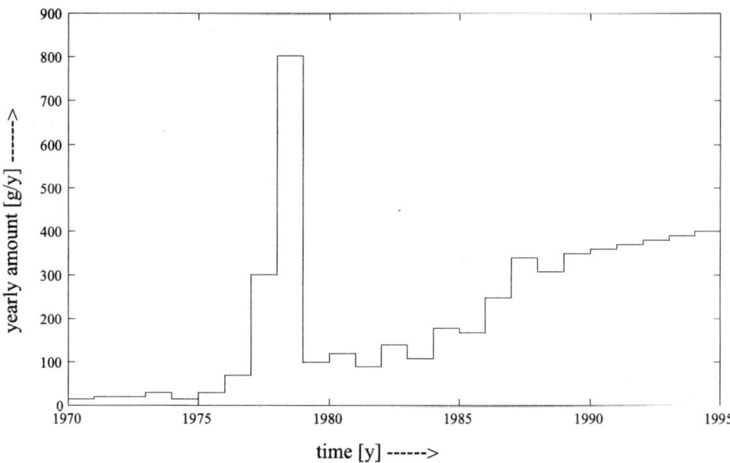

Figure 1.1 Historical development of worldwide civilian demand for tritium. Values for 1991 to 1995 are estimates.

after concerns about radiological problems arose. Since then, the yearly tritium demand slightly increased again to about 400 g/y at the beginning of the 1990s. The international trade in the 1980s was on average slightly more than 220 g per year.[5]

Demands for fusion research have increased their share to about 10%. Shipments of more than 1 g are received from manufacturers at most two to five times a year. A single production charge for luminous paints contains a maximum of 0.1 g of tritium. Only 4 out of 21 large commercial tritium manufacturing and trading facilities are located in the nonnuclear-weapons states, Canada, Germany, and Switzerland (see Appendix A on tritium facilities).

Nearly all civilian demand was satisfied by supplies from the ORNL sales office.[6] Its price has risen from $10,000 per gram in the early 1980s to $29,000 in the late 1980s.[7] It is conceivable that China, the U.K.,[8] France, and Russia[9] would be able to export tritium originating from their military production in significant quantities (several tens of grams). Smaller amounts of tritium are available from civilian sources in Belgium[10] and France.[11] In the early 1990s, Ontario Hydro (Canada) appeared as the main supplier of tritium from civilian sources on the world market. This company initially intended to gain approximately 2.5 kg of tritium per year from its newly built Tritium Removal Facility (TRF) at Darlington. It started to undertake a special effort with the support of the Canadian government to establish new applications of tritium (especially runway landing lights) to increase the market and to sell as much as possible of its tritium. The rest of the material is left to decay.

The main importing countries are the U.K., Japan, and Switzerland. The stated uses are basically for self-powered lights and radio-luminous paint. Table 1.3 presents the variety of commercial products which make use of tritium. Most of them contain very small quantities of tritium, as can be seen from the table, which shows the different applications ordered by the decreasing quantities typically required.

Table 1.3 Tritium in commercial products.

product	content [GBq] [a]	content [μg]
radio-luminous products containing tritium in paint or plastic		
compasses	0.2–2	0.5–5
instrument dials and markers, automobile shift quadrants	1	2.5
sprit levels	0.2–1	0.5–2.5
timepieces	0.04–1	0.1–2.5
automobile lock illuminators	0.07–0.6	0.2–1.5
bell pushes, rims for underwater watches	0.01	0.03
radio-luminous products containing tritium in sealed tubes [b]		
runway landing lights for remote airfields	<40,000	<100,000
exit signs	70–4000	200–10,000
helicopter rotor blade tip markers	2000	5000
night sight for artillery weapons	70–2000	200–5000
aircraft instrument panel illumination	70–700	200–2000
map readers for night use	200	500
markers	150	250
instruments, signs and indicators, step markers, mooring buoys and lights	70	200
marine compasses	7–70	20–200
public telephone dials	20	50
night sights for small arms (pistols, rifles)	4–20	10–50
timepieces, ordinary compasses	7–15	20–40
backlights for liquid crystal displays in watches, light switch markers	7	20
bell pushes	0.4	1
electronic and other devices		
tritium targets for neutron source, typical size	0.04–100	0.1–300
tritium targets for neutron source, large special	<220,000	<615,000
gas chromatographs	10	25
electronic tubes	0.0001–0.4	0.0003–1
antistatic devices contained in precision balances	0.04	0.1
cold-cathode tubes	0.003	0.008
glow lamps	0.0004	0.001
low-energy electric lights	0.2×10^{-6}	0.5×10^{-6}

[a] Figures are taken from International Atomic Energy Agency (1979) and other sources.
[b] These products are sometimes called "beta lights" or gaseous tritium light sources (GTLSs).

There are also some applications of tritium in research (Evans, 1974). Most research applications require only very small quantities (a few μg or less). In the following list the various applications are arranged in order of decreasing requirements:

- Fusion energy research (presently a few 100 g worldwide)[12]

  In recent years, fusion research makes increasing use of tritium. The main goal is to learn about the behavior of tritium in various components of fusion reactors, such as breeder blankets, off-gas processing systems, structure materials, etc. If a fusion power reactor ever started operation, it would contain a steady inventory of a few kg and burn about 180 kg per $GW_e y$.

- 14 MeV neutron sources (up to 600 mg each) (Kobisk, 1989)

  Tritium can be used for relatively compact generators of 14 MeV neutrons.[13] These high energy neutrons are useful for various applications such as neutron activation analysis, production of tiny amounts of special radionuclides, and for the study of nuclear reactions induced by 14 MeV neutrons.

- Biological, medical, chemical, and geological research, especially as tracer (μg quantities)

  Tritium is frequently used as tracer for chemical compounds which contain hydrogen. Some of the hydrogen is replaced by tritium. The behavior of the labeled molecules (pathways, residence time, etc.) can be studied by measuring the activity of different samples. Tracer studies are conducted especially in biology, medicine, physics, chemistry, agriculture, and geology. Small amounts of tritium are used in experiments as a source of Bremsstrahlung or beta particles.

1.3.2 Military uses of tritium

There are various military applications of tritium within the category of conventional weapons. These are primarily dual-use applications, mostly using tritium as self-powered light sources for conventional weapon systems. They are similar to commercial applications of tritium and are included in Table 1.3. The largest quantities of tritium for these applications are used in airfield runway landing lights like those employed during the U.S. invasion of Grenada.

Some tritium is used for research in inertial confinement fusion (ICF). These experiments are closely linked to nuclear weapons research (Schaper, 1991 and Gsponer and Hurni, 1998).

The largest amount of tritium required for military purposes is used in nuclear weapons. In these, tritium is mainly used to generate neutrons from fusion with deuterium according to the following equation:

$$D + T \rightarrow \alpha + n + 17.6\,\text{MeV}. \tag{1.4}$$

The fusion energy released by this reaction does not contribute significantly to the total yield of the nuclear weapon.

Although attempts have been made to construct thermonuclear weapons which do not require a fission primary, it is generally believed that they have not been successful. Some analysts (Gsponer and Hurni, 1998; Makhijani and Zerriffi, 1998; Jones and von Hippel, 1998; and Jones *et al.*, 1998) fear that this might change with the increased research efforts regarding thermonuclear weapons physics (U.S. Department of Energy, 1996). Thus, tritium is neither sufficient nor necessary to produce a nuclear weapon. Obviously, it is less significant for nuclear weapons than uranium and plutonium. However, it becomes strategically significant as an addition to fissionable material. Tritium is believed to be a component of most nuclear weapons currently in the stockpiles of all nuclear weapons states. Its primary purpose is to increase ("boost") the explosive yield of a given amount of fissionable material. Hence, the particular significance of tritium for the nuclear arsenal is to guarantee a higher total yield. Without tritium the total yield would be lower by one order of magnitude (see Sections 1.6.2 and 1.6.3).

Tritium is further used for selectable yields, for radiation enhanced weapons, and for the thermonuclear stage in fusion weapons. The use of tritium in nuclear weapons, therefore, leads to vertical proliferation from first-generation fission devices to more sophisticated boosted or thermonuclear weapons.

The following list explains the different uses and estimated quantities of tritium in nuclear warheads:[14]

1. *Neutron generator (0.1 mg):* The nuclear fission chain reaction is started by a shower of neutrons. In early weapon designs a bimetallic neutron source was used for this purpose. Pulverized beryllium and an alpha emitter are mixed and generate neutrons according to the nuclear reaction $^{9}\text{Be} + \alpha \rightarrow {}^{12}\text{C} + \text{n}$. A Ra-Be neutron source with 10 Ci Ra-226 emits some 10^8 neutrons per second (see, e.g., Lieser, 1980). Within the first generation of neutrons[15] this source emits just one neutron. A stronger pulse of neutrons is desirable because the higher the number of neutrons in the first generation, the more reliably can the beginning of the chain reaction be predicted.[16] Precise timing of the pulse is more important, however, as it has a considerable impact on the yield.

   This can be achieved with an electrostatic neutron generator in which deuteron ions accelerated by a high voltage strike a metal tritide target.[17] Typical targets contain up to 0.1 mg of tritium (see, e.g., Evans, 1974). Neutron showers with some 5×10^{12} neutrons per second are available for civilian purposes (Lieser, 1980). Such neutron generators were first used by the U.S. in 1953 and are today as small as a fist (Cochran, 1987a).

   In contrast to the bimetallic neutron source, this type of neutron generator is an external neutron source, i.e., neutrons are shot from the outside into the core of the critical mass. The cavity in the core is a precondition for the boosting which is described in the next paragraph.

2. *Boosted fission weapons and boosted primaries in thermonuclear weapons (2–3 g):*[18] The principle of fusion-boosted fission bombs was declassified in 1974.[19] In these devices a mixture of tritium and deuterium gas is injected from an external gas capsule at high pressure (about 10^7 Pa) into the hollow core of the plutonium sphere before the chain reaction is initiated. For a short time, both temperature and pressure in the center of the nuclear explosion are high enough to allow tritium and deuterium to fuse. The neutrons released

from this fusion induce further fissions, thereby increasing the efficiency of the fissionable core and possibly of the surrounding tamper (fertile material, e.g., uranium-238). As a result, the yield can be multiplied — depending on the fission efficiency without boosting — by a factor of 2 to 10 (Hansen, 1988) or even 100.[20]

As a consequence, boosted weapons can achieve explosive yields up to 400 kt,[21] but are still heavy. However, thermonuclear weapons with boosted primaries can have high yield-to-weight ratios (see Figure 1.4), allowing high-yield (100 to 500 kt) warheads to be small and light enough (100 to 400 kg) to fit into long-range missiles with multiple warheads, as well as into torpedoes and artillery shells (see Figure 1.4).

In addition, it may be assumed that boosted weapons are more reliable than ordinary fission bombs concerning their explosive yield, because the reduced quantity of plutonium is easier to compress by means of a conventional explosion and thus has a reduced probability of predetonation. To put it another way, the probability of predetonation remains the same when a plutonium bomb with a lower content of plutonium-239 utilizes tritium.

Boosting is believed to be applied in most of the small fission weapons and in all triggers of thermonuclear weapons in the current U.S. nuclear arsenal (see Section 1.6.2).

The quantity of tritium used for boosting (2–3 g) can be estimated from the U.S. stockpile on the assumption that all nuclear weapons make use of boosting and that not all military tritium is used for this purpose.[22] The complete fusion of 3 g of tritium with 2 g of deuterium releases enough neutrons to induce fission of 240 g of plutonium. The fission releases more than 4 kt of TNT, whereas the fusion energy contributes no more than 0.4 kt. In thermonuclear weapons the largest fraction of the energy comes from fusion (see item 5 below).

A high purity is required for the mixture of tritium and deuterium.[23]

3. *Selectable yield (2–3 g):* This is a special feature used for some boosted warheads. The desired yield can be selected immediately before its delivery by varying the timing of its application or by inserting more or less (from zero to all) capsules each containing a fraction of the total tritium. There are some other methods that can be used as well to enable yield selection. The new modification of the B61 (mod. 11) has the widest known range of yields from 0.3 to 340 kt (Mello, 1997).

4. *Neutron bomb (Gsponer, 1984 and Huaqiu, 1988) (10–30 g):*[24] Tritium is used in a similar way in radiation enhanced weapons (neutron bombs), which are designed to maximize the portion of energy which is carried by the neutrons. The 14 MeV neutrons generated by fusion have a higher probability of escaping than of being stopped by inducing further fissions. Therefore, these weapons have a comparatively low yield (about 1 kt of TNT) and a large flux of high-energy neutrons. The amount of tritium needed for this sort of nuclear weapons is of the order of a few tens of grams.

   When no tritium is inserted, a neutron bomb works like a simple low-yield fission weapon.

5. *Thermonuclear weapon:* The primary material of thermonuclear stages is LiD. Tritium is bred in situ from lithium by capturing a neutron. Some of the lithium hydride can be added in the form of LiT. This is not necessary, but is probably done in some weapon designs to facilitate the beginning of the fusion reaction (see, e.g., Winterberg, 1981). Quantities are not publicly known.

Most information regarding the use of tritium for nuclear weapons is classified. More details are described elsewhere (see, e.g., Albright and Paine, 1988; Cochran *et al.*, 1984; Gsponer and Hurni, 1998; Hansen, 1988; Seifritz, 1984; and Winterberg, 1981). All quantities given above are educated guesses which have been neither confirmed nor denied by official sources.

The strategic significance of tritium for the nuclear arsenal is to provide a high total yield or a high yield-to-weight ratio (see Section 1.6.2).

The military facilities for the production of tritium and their capacities in the nuclear weapons states are described in Section 2.7. Tritium-related activities in de facto nuclear weapon states are dealt with in Section 1.7.

1.3.3 Civilian/military ambivalence of tritium

Tritium has various military (nuclear and nonnuclear-weapons related) as well as civilian (industrial and scientific) applications.[25] The relationship between civilian and military uses of tritium becomes particularly apparent when reviewing the historical development.

History of civilian and military uses of tritium:[26]

1934 The British scientist Lord Rutherford discovers tritium by a nuclear reaction.

1939 Luis Alvarez from the University of California discovers the radioactive decay of tritium.

1944 During World War II, the first calculations regarding the use of fusion energy in nuclear weapons are undertaken by scientists of the Manhattan Project in Los Alamos.

1950 Under the impression that the first Soviet nuclear weapon test has been carried out, research on thermonuclear weapons is enhanced in the U.S.

1951 On May 8, tritium is used for the first time in a nuclear weapon test by the U.S. During "Operation Greenhouse," a fusion-boosted fission device with the code name "George" is tested on the Enewetak Atoll in the West Pacific.

1952 On November 1, the U.S. explodes the first thermonuclear device at the same site. This shot, called "Mike," weighs 82 tonnes and has a yield of 10.4 Mt of TNT, that is, 500 times the yield of the Nagasaki bomb.

1955 At the Savannah River Plant (South Carolina, U.S.), the first military reactor dedicated to the production of tritium starts operation. Further military tritium production reactors are commissioned at Marcoule (France, 1956), Chelyabinsk (U.S.S.R., 1957), Chapel Cross (U.K., 1958), China (1968) and Dimona (Israel, 1968).

1958 The strict classification of information about tritium and fusion research is relieved. The magnetic fusion project of the U.S. is declassified at the second conference on Atoms for Peace in Geneva. Subsequent declassifications show that France and the Soviet Union have made progress in fusion research which is comparable to that of the U.S. (Office of Technology Assessment, U.S. Congress, 1987).

1959 The U.S. Atomic Energy Commission (USAEC) makes 100 g tritium available for peaceful applications. In comparison, 4.1 g of tritium had been sold in 1958, and only 1.3 g in the period between 1948 and 1957 (Anonymous, 1959).

1961 An international conference in Vienna on applications of tritium in physics and biology is attended by 290 scientists from 27 countries (IAEA, 1962).

Early 1960s Tritium is released regularly from U.S. military production for worldwide civilian uses. This pushes industrial application on a large scale, especially the use of tritium for radio-luminous paint.

1964 The ENEA (European Nuclear Energy Agency of the OECD) and the IAEA constitute a common commission for the elaboration of recommendations on radio-luminous timepieces.

1966 According to a recommendation by the ENEA and IAEA, radium in timepieces should be replaced by the radiologically less dangerous tritium or promethium-147 (IAEA, 1967). Other technically possible alternatives such as carbon-14 and nickel-63 are assessed as feasible options, but are not recommended because they are not "commercially" available.

1967 During negotiations on the Nonproliferation Treaty (NPT), it is decided not to define tritium as "special nuclear material" and not to place it under international control.

1974 Declassification of the use of tritium in nuclear weapons by the U.S.

1978 The company American Atomics is closed because radiation protection regulations were violated when producing radio-luminous paint.

1980 The civilian use of tritium sees a deep decline when radiation protection standards are tightened.

1985–1986 Pakistan illegally receives 0.8 g of tritium, certain handling instruments for tritium, as well as technical advice from the German company NTG.

1986 Tritium and tritium technology are included in the Atomic Energy List (Group 2) of CoCom. This is the first step towards tritium control on the international level.

1988 Shutdown of all military production facilities for tritium in the U.S. in April. For the first time a verified production cutoff for tritium is publicly discussed.

1989 U.S. Congressman Edward Markey introduces a bill (H.R. 2502) to ban all tritium production for nuclear weapons at commercial facilities. This ban would be lifted in the case of war or national emergency. The bill is not passed.

1989 Ontario Hydro (Canada) decides that it will offer the output of its detritiation facility at Darlington on the market.

1989 On August 30, the Canadian government gives its approval to Ontario Hydro to export tritium for specific uses and under strict control.

1990 In September, the 4th NPT review conference recognizes that tritium is relevant to the proliferation of nuclear weapons and therefore calls for "early consultations among countries to ensure that their supply and export controls are appropriately coordinated" (Nonproliferation Treaty, 1990).

1990 Ontario Hydro announces the start-up of the Darlington detritiation facility after one year of abortive attempts. It is by far the largest civilian source for tritium able to recover some 2.5 kg of pure tritium gas per year.

1991 In May, the extension of the cooperation agreement between EURATOM and Canada is finalized (EURATOM and Canada, 1991). EURATOM will act as a supervising agency authorized to establish control procedures for tritium supplied from Canada to fusion research facilities in EURATOM member states.

1992 A new dual-use list is adopted at the meeting of the 27 adherents to the Nuclear Suppliers Guidelines in April in Warsaw. These new guidelines cover not only tritium, tritium compounds, and mixtures, but also tritium facilities or plants and components (Nuclear Suppliers Group [NSG], 1992).

1993 In December, the United Nations General Assembly adopts a resolution calling for negotiations on banning the production of nuclear materials, but does not specify the substances to be covered. Tritium may be included, but is scarcely mentioned in the discussions.

1994 The Gore–Chernomyrdin agreement on the verified cutoff of plutonium production reached in June is an important bilateral step by the U.S. and Russia towards a fissile materials production cutoff. However, three tritium production reactors (2 in Russia, 1 in the U.S.) are excluded from the agreement, although plutonium could easily be produced in them.

1994 In September, the proposal to expand international nuclear safeguards to cover tritium is presented at the IAEA Symposium on International Safeguards in Vienna (Colschen and Kalinowski, 1994). It faces some skepticism regarding its technical feasibility and political desirability.

1995 A workshop, entitled "Fissile materials and tritium — How to verify a comprehensive production cutoff and safeguard all stocks," took place at the Palais des Nations in Geneva on 29/30 June.[27] It became apparent that it is necessary to consider ongoing tritium production in any attempts to ban the production of plutonium. Various proposals to put tritium under international control were discussed and further developed. A new idea that came up there is that tritium control might play a key role in getting nuclear arms control started on the South Indian subcontinent (Hoodbhoy and Kalinowski, 1996).

1995 In October, the U.S. Department of Energy releases the "Final Programmatic Environmental Impact Statement for Tritium Supply and Recycling." A dual-track strategy is proposed for developing a reliable source of tritium for military purposes. One track is to utilize a commercial reactor; the second is based on a linear proton accelerator (U.S. Department of Energy, 1996).

1996 In October, it becomes public that the private U.S. company Advanced Nuclear and Medical Systems (ANMS) suggests using German plutonium as fuel for the Fast Flux Test Facility (FFTF) in Hanford and using this facility to produce tritium for U.S. nuclear weapons.

1998 In August, the Conference on Disarmament agrees to establish an Ad Hoc Committee to negotiate a "Treaty Banning the Production of Fissile Material for Nuclear Weapons or Other Nuclear Explosive Devices" (Cutoff Treaty).

1998 At the end of the year, the U.S. Department of Energy selected commercial power reactors as the primary method to produce tritium for military purposes. The accelerator-based method will comprise an assured backup capability.

2003 In September, the breeding of tritium for military purposes is resumed by the U.S. at the Watts Bar pressurized water reactor.

2005 After an 18 months irradiation period the first batch of tritium will be extracted by the U.S. at the Savannah River site.

The chronology listed above shows that a breakthrough for commercial applications of tritium was made possible in the early 1960s by the release of significant amounts of tritium from military production, beginning just a few years after production has started. Civilian applications are still dwarfed by the quantities used in nuclear weapons. Figure 1.2 gives a comparison of annual inventory changes.[28] It demonstrates that annual civilian uses are still much lower than annual military inventory changes.

The relevance of tritium for proliferation has long been underestimated due to the late declassification of its use in nuclear weapons. During the negotiations for the Nonproliferation Treaty (NPT) the significance of tritium for nuclear weapons was not generally known.[29] It took nearly 30 years from the first release of tritium from military production to civilian use before the relevance of tritium to proliferation was recognized at the 4th NPT Review Conference in 1990. Only in the past few years have international efforts to control the proliferation of tritium for nuclear weapons been taken seriously. Nevertheless, it remains a piecemeal approach with significant loopholes, and a coherent international strategy has yet to be devised. Even within the international scientific community engaged in tritium research, the military relevance is not adequately recognized.[30]

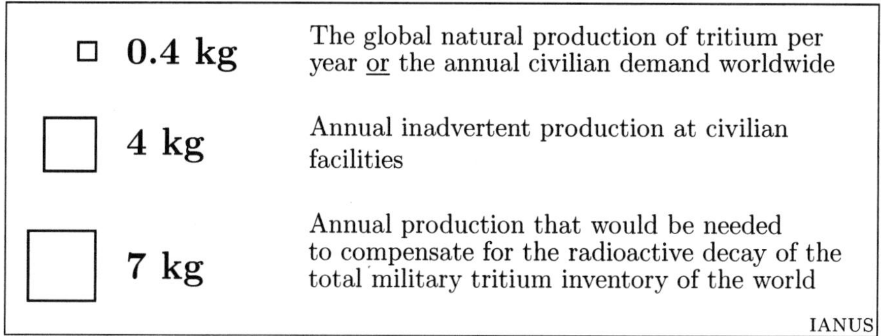

Figure 1.2 Comparison of different rates of tritium production and decay. The yearly civilian demand for tritium by the industrialized countries is around 0.4 kg. The natural production rate in the ecosphere is estimated to be of the same order of magnitude. The inadvertent production in civilian facilities amounts to (4.1±1.0) kg/y. The annual decay of tritium in all nuclear weapons of the world, including those withdrawn from the operational stockpiles, is estimated to be (6.8±2.0) kg at the end of 1998.

The quantity, mode of production, physical condition, or degree of purity of tritium by no means determines the intended military or civilian use of tritium.[31] Tritium produced in military facilities is neither technically nor politically confined to military uses. On the other hand, the largest civilian producer of tritium, Canada, tries to set legal boundaries to ensure that civilian-produced tritium can be used for civilian purposes only. The U.S. follows a comparable, although not legally binding, policy regarding tritium originating in civilian facilities. This is seriously counteracted by the dual-track program of the U.S. Department of Energy as decided in 1995, which includes the strategy of purchasing irradiating services or an existing commercial power reactor for the production of tritium for military purposes (U.S. Department of Energy, 1996). Even if the linear accelerator based system is selected as the primary method for tritium production, the commercial path will first be used on an interim basis and then remain as a fully proven backup capability.

Since it is impossible to physically differentiate between "military tritium" and "civilian tritium," the respective sociotechnological environment has to be taken into account when a judgment is required. Any quantity of tritium can become militarily significant. Since physical barriers can never be completely tight, the most efficient way to prevent tritium diversion for military purposes — in addition to binding and verified political commitments — is to **minimize any production and application** and to follow the most **proliferation-resistant path** for those applications which cannot be avoided.

Many applications which were introduced more or less because tritium became available at low cost can be completely abolished or replaced by nonradioactive technological alternatives. Radiation protection can serve as a driving argument for such a conversion.[32]

For most civilian applications of beta emitters, tritium can be replaced by other isotopes such as carbon-14, nickel-63, krypton-85, promethium-147, or thallium-204 (see, e.g., Heusinger and Rau, 1961). In particular, promethium-147

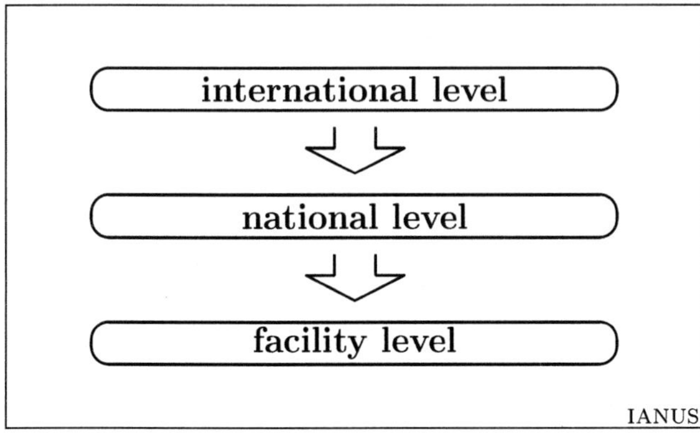

Figure 1.3 Levels of tritium control.

has been recommended as an alternative to tritium in radio-luminous paint. The isotopes carbon-14 and nickel-63 have been assessed as feasible options, but were not recommended because they were not "commercially" available (IAEA, 1967). Tritium cannot be replaced by other beta-emitters if the chemical behavior of hydrogen is required. Therefore, tritium is, for example, required for tracer studies of hydrogen and compounds containing hydrogen which cannot be labeled with other isotopes.

Another example concerns a technology which requires handling of tritium that has been produced inadvertently in the moderator of heavy water reactors. For radiation protection purposes it is desirable to remove the tritium from the heavy water. The conventional method used for this purpose extracts tritium gas of high purity. The more proliferation-resistant alternative would produce an off-stream which contains tritium in aqueous form and enriched at less than 30%. This can then be further treated as aqueous radioactive waste and conditioned with concrete to be enclosed in a solid matrix.

1.4 Current state of tritium control

1.4.1 Tritium control at the facility level

Three levels of control can be distinguished and will be discussed here: the facility, the national, and the international level (see Figure 1.3). These three levels are interrelated and hierarchically ordered. In general, measures taken on a higher level have effects on a lower level. Tritium control at these three levels has developed to a very different extent. In all countries, some sort of tritium control for radiation protection purposes is obligatory at the facility level. Most countries have national control systems which become effective above a certain licensing limit. But at the international level, tritium control is still in its infancy.

Tritium control is common practice at all tritium handling facilities for radiation protection purposes. A quite elaborate system has been established at the laboratories run by the U.S. Department of Energy (DOE). The accountability procedures practiced by the Tritium Research Laboratory (TRL) at *Sandia National Laboratories Livermore (SNLL)* can serve as an example (Wall and Cruz, 1985).

The Nuclear Materials Management Group (NMMG), which is responsible for the accounting of nuclear materials to the DOE, has divided SNLL into Material Balance Areas (MBAs) with a custodian assigned to each MBA. The NMMG requires that the quantity of material be verified and that any discrepancies in shipping information be resolved before the tritium is transferred to the MBA. When the receipt procedure is complete and any shipping discrepancies are resolved, the NMMG issues the tritium to the MBA custodian through the use of the Nuclear Material Control Transfer Slip. The custodian in turn issues the material to the personnel of his or her respective MBA. The personnel then becomes accountable to the MBA custodian for that nuclear material. All shipments of materials containing tritium are to be processed through the NMMG. Hand carrying of tritium or its transport in personal vehicles is prohibited.

The quantity of tritium assigned to a container is recorded to the nearest milligram (0.001 g, 0.36 TBq). All quantities of tritium equal to or greater than 0.0005 g (0.18 TBq) are considered accountable within one MBA.[33]

1.4.2 Tritium control at the national level

State's system of accounting and control (SSAC)

Organizational arrangements at the national level to account for and control tritium would provide the essential basis as well as a reinforcement for the application of international agreements regarding the control of tritium. Most of the current tritium supplier countries have implemented organizational arrangements on a national level to account for and control the import, export, and handling of radioactive substances including tritium.

General regulations regarding the handling of tritium include licenses for the construction of tritium-handling facilities; for the possession, production, transportation, manufacturing, and marketing of tritium and tritium-containing products; and for the design of specific consumer products, as well as for waste disposal.

Reasons for the control of tritium are

1. public as well as occupational health and radiological safety
2. environmental protection
3. minimization of material loss for economic reasons (tritium is very expensive)
4. prevention of nuclear proliferation

Mainly regulations concerning the last goal are reviewed here.

Control of tritium in Germany

The relevant law for the acquisition/possession and production of "other radioactive substances" is the Radiation Protection Ordinance (Strahlenschutzverordnung, StrlSchV). The limits above which a license is required and the appropriate exemption limits are regulated by §3 and §4 in combination with the appendix to the

Radiation Protection Ordinance.[34] Possession must be reported to the responsible authority above 5×10^6 Bq, and above 5×10^7 Bq a license is required.

Exemptions from these provisions are possible under certain conditions (StrlSchV, Appendices II and III).[35]

Both the import and the export of radioactive substances requires a license or a declaration (StrlSchV, §11–14; Atomic Energy Act = Atomgesetz, AtG, §3). The licensing authority is the export control agency (Bundesausfuhramt).

The Atomic Energy Act and the Radiation Protection Ordinance expressly stipulate that other legal provisions on import and export remain unaffected. In this context, the Foreign Trade and Payments Act and the related Foreign Trade and Payments Regulations (Außenwirtschaftsverordnung – AWV) as well as the War Weapons Control Act (Kriegswaffenkontrollgesetz) are of particular relevance.

The latter restricts the export of any weapons of war. The definition of weapons of war can be found in the Weapons of War List. According to this list (Part A, point I.2) "any part, device, assembly or material especially designed for, primarily useful in" nuclear weapons are treated as weapons of war. The nuclear fuels are further defined by certain fissionable isotopes and "any other material capable of releasing substantial quantities of atomic energy through nuclear fission or fusion or other nuclear reaction of the material. The foregoing material shall be considered to be nuclear fuel regardless of the chemical or physical form in which they exist."

The Federal Court of Justice (Bundesgerichtshof) passed a sentence (see below and Bundesgerichtshof, 1992) according to which tritium is "primarily useful" for nuclear weapons irrespective of its amount. Thus, any unlicensed tritium export is considered an export of a weapon of war, even if it is destined for a declared civilian purpose.

Annexed to the AWV is a list of items which require an export license (Ausfuhrliste), Part I B of which forms the Nuclear Energy List (Kernenergieliste), which corresponds to the atomic energy list of CoCom that was replaced by the Wassenaar Agreement in 1995. A number of restrictions apply to trade with these "sensitive" nuclear materials, installations, and equipments which are defined in combination with the country lists C and H (for a detailed description see Müller *et al.*, 1994).

The Nuclear Energy List covers tritium in the following way:

> 0112: Tritium and compounds, mixtures and products containing more than 0.1 per cent tritium as well as products containing one or more of the aforesaid substances except:
> i) consignments with an activity below 100 Ci
> ii) tritium contained in luminous paints, self-luminous products, gas and aerosol detectors, electron tubes, lightning or static elimination devices, ion-generating tubes, detector cells of gas chromatography devices, and calibration standards
> iii) compounds and mixtures of tritium where the separation of the constituents cannot result in the evolution of an isotopic mixture of hydrogen in which the tritium portion exceeds 0.1%
>
> 0305: Equipment especially designed for tritium production or recovery.

A declaration about the end-use (Endverbleibserklärung) is obligatory, but no control measures are in force.

In 1986 and 1987, the German company NTG illegally exported tritium and tritium-handling facilities to Pakistan. Three individuals were found guilty and sentenced to fines and prison, some of them on parole. One of them appealed against his conviction. But the responsible court of appeal (Federal Court of Justice, Bundesgerichtshof) approved the sentence with the argument that tritium would in any case be covered by the Weapons of War Control Act (Kriegswaffenkontrollgesetz) irrespective of its quantity (Bundesgerichtshof, 1992). Even the smallest quantity was to be regarded as a material usable for nuclear weapons and therefore required an export license. An export to a country suspected of running a clandestine nuclear weapons program would not be licensed and would represent a violation of the Weapons of War Control Act even if the tritium was not used directly for a nuclear weapon.

Control of tritium in the U.S.

Obviously the U.S. has a state's system of accounting and control (SSAC) both in the civilian and in the military field, and it is likely that the same is true for other nuclear weapons states. In the U.S., a license can be issued to companies that produce tritium-containing products. They have to prove that they fulfill several obligations for radiation protection. In the application form, the manufacturer has to indicate the supplier of the tritium, the dates of the first and final shipments, and the method, as well as the quantity of each transport. Any licensee must submit an annual report indicating the number of products manufactured during the year and the quantity of tritium contained. If a licensee has more than 1 g of tritium in stock at any given time and in any location, an inventory statement has to be submitted twice a year. It is obligatory to provide immediate notification if any attempted or apparent theft of tritium has occurred.

Some specific information is available on the system of tritium accounting and control established by the U.S. Department of Energy (DOE) (Wall and Cruz, 1985). The purpose of the system is to assure facility managers and DOE officials that no material has been lost either in the system or to the environment, and that no tritium was diverted for any use besides the one originally intended.

All physical and chemical forms of tritium (including scrap[36]) are accountable material. Each facility run by the DOE has its own internal system of accounting and control for tritium. Periodically, the tritium within the facilities must be physically inventoried, assayed, and reported to the DOE. For this purpose, the quantities are summed up to the nearest mg. The accountable quantity of tritium is always rounded to the nearest 0.01 g (3.6 TBq) for DOE reporting purposes. Quantities less than 0.005 g (1.8 TBq) are rounded to 0.00 and are not considered accountable to the DOE. Any tritium acquired for projects must be authorized by the DOE (Wall and Cruz, 1985).

For special nuclear materials (SNM), the DOE requires measurement control programs. Although tritium is not an SNM, such a program may be conducted because of its strategic and economic value.

According to the Code of Federal Regulations, Title 10, Chapter 1, §110.23 (a) 2, commercial exports above 10 Ci (=0.37 TBq $\simeq$1 mg) in any dispersed form need a license from the Nuclear Regulatory Commission (NRC).

Discussions on export control in the U.S.

In the wake of discussions on the fate of military production reactors, different groups in the U.S. raised concerns about using tritium from civilian sources for U.S. nuclear weapons (see, e.g., Markey, 1989). In 1989, U.S. Congressman Edward Markey introduced a bill (H.R. 2502) to ban all tritium production for nuclear weapons at commercial facilities. This ban would be lifted in the case of war or national emergency. The bill was not passed.

At the same time, the possible horizontal proliferation of tritium gained increasing public attention. The Nuclear Regulatory Commission (NRC) believed that its current export licensing procedures were adequate because tritium was considered of less strategic significance than special nuclear materials. Nevertheless, NRC admitted that it would be preferable if assurances regarding the prevention of unauthorized retransfers of U.S.-supplied bulk tritium were given in written form (Zuercher, 1991).

On June 25, 1991, the Nuclear Proliferation Prevention Act was introduced in the U.S. House of Representatives. Its purpose is to restrict the export of dual-use items and to promote negotiations with other nations to make changes in the IAEA's safeguards effort. The negotiations would aim at the application of IAEA safeguards to tritium (Hibbs, 1991).

As far as the regulations for the control of tritium at the facility level are concerned, there are numerous reports that shipper–receiver discrepancies occurred (see Section 3.4.3). In one case the discrepancy was as large as 40% of the original order.[37] In another case 2.2 grams were missing after an internal transfer of tritium from one building to another within Oak Ridge National Laboratory.

The DOE procedure following those cases has usually been a suspension of further commercial tritium sales and the initiation of a subsequent investigation. These investigations by the DOE, which were supported by the NRC, since these transactions legally require NRC export licenses, resulted in statements that the discrepancies were not due to leakage, procedural problems, theft, or diversion, but to management methods and the notoriously difficult problem of containing the gas. However, a report of the DOE's inspector general published in October 1989 stated that three investigations of the problem had failed to adequately weigh the possibility of theft or diversion and posed a potential concern with respect to safety as well as nonproliferation. These doubts did not lead to the stopping of commercial tritium sales by the ORNL. Sales have been resumed despite the failure to determine the whereabouts of the missing tritium. The only attempt to correct the shipping problems was a change in measurement methods (the use of calorimetric measurements) by the ORNL for its tritium shipments and calls for more effective management oversight of the entire commercial sales program. The existing regulations were found to be appropriate.

A report published by the U.S. General Accounting Office (GAO) in March 1991 came to the conclusion that controls over commercial sales and export of tritium could be improved (U.S. General Accounting Office, 1991). The report revealed severe weaknesses in tritium accounting procedures at Oak Ridge National Laboratory and a lack of oversight and inventory control practices by the Department of Energy. The GAO recommended that the Nuclear Regulatory Commission should require end-use statements and obtain written agreements from recipient countries for notification/approval of retransfer of exported U.S. tritium.

Discussions on export control in Canada

Canada has become the world's largest civilian producer of tritium. The 16 operating CANDU-type heavy water reactors with a total of 12,000 MW_e inadvertently produce up to about 3 kg of tritium per year in their moderator and coolant. The company Ontario Hydro decided to build a facility for the detritiation of the heavy water. The extraction of tritium from heavy water is undertaken primarily to reduce emissions and to minimize the occupational radiation dosage, a quarter of which is caused by tritium. The Tritium Removal Facility (TRF) at Darlington is designed to extract about 2.5 kg per year, which is worth more than $25 million. Therefore, detritiation for radiation protection purposes has the positive side-effect that it yields a precious isotope of a high purity and in large quantities.[38]

During the construction phase (end of the 1980s), a public discussion with broad media coverage took place. Its main focus was the question of what should be done with the tritium and how to avoid its use for nuclear weapons. The focus was on the public demand that Canadian tritium must not be used in U.S. nuclear weapons and should not free U.S. military production capacities by satisfying the needs of commercial customers that used to be served by the U.S. ("substitution argument"). Some groups such as Energy Probe even suggested storing the tritium safely or dumping it in an appropriate nuclear waste disposal and letting it decay.

In 1985, the so-called Tritium Issue Working Group was formed and prepared an extensive report to investigate this question (Spratt, 1985). It came to the conclusion that the current Canadian export control legislation in combination with a careful licensing policy to be adopted by the Atomic Energy Control Board (AECB) would suffice to satisfy concerns about the nuclear weapons use of tritium. It recommended not to designate tritium a safeguardable product because this would "destroy any potential commercial opportunities that exist in this area for Canada" (Spratt, 1985, p XV).

An expertise by the Canadian Environmental Law Association prepared in 1986 came to the conclusion that the regulations on use, sale, and export of tritium do not effectively guarantee that Ontario Hydro's tritium is used for peaceful purposes only. The lawyers recommended that the IAEA and the NPT be amended to include tritium and that trilateral safeguards agreements be negotiated between Canada, the IAEA, and the prospective purchasers (Canadian Environmental Law Association, 1986).

In 1989, Ontario Hydro eventually decided, after a thorough investigation that it would offer the tritium on the market. Ontario Hydro and the Canadian government were aware of the proliferation risks of tritium exports. Both reacted to the public pressure by adopting a special "tritium nonproliferation policy" which intends to guarantee that Ontario Hydro's tritium is used for peaceful purposes only. On August 30, 1989, Canada's Energy Minister Lyn McLeod approved limited export of tritium for specific uses under additional precautions, including special contractual obligations, detailed documentation, special transportation procedures, on-site inspections, and audits (Ontario, 1989).

In any case, tritium is covered by the Canadian export control legislation. The two relevant federal instruments are the Export/Import Permits Act and the Atomic Energy Control Act. The AECB is the authority responsible for issuing export licenses. In March 1986, this authority released guidelines for the export of tritium (Sinden, 1986). The issue of permits for the export of tritium is subject to the specific conditions and limits on quantity for each state in any year. The limits depend on whether the receiving state is a member of the NPT or "in good standing." For the latter, the export limit for pure tritium has been set at 370 TBq, which is the highest limit value of all countries (see Table 1.4).

The Canadian legislation does not allow any tritium sales to nuclear weapons programs of the five recognized nuclear weapons states. In order to prevent the horizontal proliferation of tritium to nonnuclear-weapons states, Canadian legislation requires in most cases a specification of the end-use and makes the license subject to a political assessment. Apart from one exception described in the next paragraph, there are no provisions for verification.[39]

Most noteworthy, in May 1991 the cooperation agreement between EURATOM and Canada was extended, according to which EURATOM acts as the supervising agency authorized to establish control procedures for tritium shipments from Canada to EURATOM member states (see Section 1.4.3).

Comparison of national export regulations

A comparative study of national regulations of accounting for and control of tritium undertaken in 1990 revealed wide disparities in national measures and the insufficiency of international cooperation to prevent the diversion of tritium for nuclear weapons purposes (Colschen, 1991).[40] The requirements for export licenses differed drastically from country to country, and no provisions to verify the end-use were made.

As an example, Table 1.4 illustrates in ascending order the broad range of national limits for license-free tritium exports, which covers more than nine orders of magnitude.[41] Some countries did not have any regulations regarding tritium; others had very weak regulations. Only among CoCom countries and the states joining the succeeding Wassenaar Agreement were the regulations somewhat harmonized.

1.4.3 Tritium control at the international level

International coordination of export controls

Nonproliferation Treaty (NPT): During negotiations of the NPT (until 1968), it was decided not to include tritium in the definition of "special nuclear materials" and not to place it under international nuclear safeguards.

In September 1990, the 4th review conference of the parties to the NPT recognized that tritium is — although not identified in NPT Article III.2 — relevant to the proliferation of nuclear weapons. Therefore, the conference called for "early consultations among states to ensure that their supply and export controls are appropriately coordinated" (Nonproliferation Treaty, 1990). No subsequent activities were initiated directly within the NPT framework, but as a consequence of this call international coordination of export controls was indeed strengthened.

Table 1.4 Maximum amount of tritium free of license requirement for export from different countries. After Colschen *et al.* (1991)

limit	country
0 GBq:	Argentina, Austria, Japan, Malaysia, Switzerland
0.0002 GBq:	Mexico
0.0037 GBq:	Finland
0.0050 GBq:	Indonesia
0.0370 GBq:	Philippines
370.0000 GBq:	U.S.
1,500.0000 GBq:	Nuclear Suppliers Guidelines, "Dual Use List" (1992)
3,700.0000 GBq:	Belgium, France, Germany, Italy, Netherlands, Norway, South Africa, U.K., CoCom (1986-1994)
37,000.0000 GBq:	Canada (to any state)
370,000.0000 GBq [a]:	Canada (to NPT members), Sweden
no limit:	ČSFR (until 1993), Hungary, Romania

[a] This is $\simeq$1 gram.

CoCom and the Wassenaar Agreement: By the mid-1980s, no multilateral agreements concerning tritium control were in force. CoCom (Coordinate Committee for Multilateral Export Control) was the first institution that coordinated national export controls of tritium.

According to CoCom regulations, licenses were required for exports of equipment specifically designed for the production or recovery of tritium and for the export of more than 3.7 TBq tritium and mixtures in which the ratio of tritium to hydrogen by atoms exceeds 1 part in 1000. Certain commercial products containing tiny amounts of tritium, such as radio-luminous paint and ion-generating tubes were exempted.

CoCom regulations did not require any provisions for verification, and not many countries adhered to the regime. The limited success can be seen from Table 1.4. Seven CoCom member states adhered to the guidelines regarding the limit for license-free export of tritium and were joined by South Africa. Three member states (Australia, Japan, and the U.S.) kept less stringent license limits.[42]

CoCom ceased to exist after 45 years following the 28–30 March 1994 meeting in The Hague (Arms Control Reporter, 1994). According to the Wassenaar Agreement, the political reform process resulted in a new control body with partly liberalized export controls which target mainly so-called "rogue states" such as Iran, Iraq, or North Korea.

Nuclear Suppliers Group (London Club): In 1991, the Nuclear Suppliers Group (NSG) was reactivated as a reaction to the second Gulf War. During their meeting on 5–7 March 1991 in The Hague, the NSG set up a working group to reinforce export controls on dual-use nuclear materials and technologies. This group elaborated an amendment of the NSG guidelines. Some countries, including Germany, suggested that tritium and tritium technology should be added to the list of dual-use items. In April 1992 in Warsaw at the next formal meeting of the adherents to the NSG, the new "Guidelines for Transfers of Nuclear-Related Dual-Use Equipment, Material and Related Technology" were adopted. Items on the new dual-use list should only be exported if certain criteria are met. In particular, the receiving country should not be on the list of proscribed countries.

The dual-use list covers not only tritium, tritium compounds, and mixtures (8.3), but also tritium facilities or plants and components thereof (8.4). These are listed under item 8, "Other" (NSG, 1992):

> 8.3. Tritium, tritium compounds, and mixtures containing tritium in which the ratio of tritium to hydrogen by atoms exceeds 1 parts in 1000 *except a product or device containing not more than 40 Ci of tritium in any chemical or physical form.*
>
> 8.4. Facilities or plants for the production, recovery, extraction, concentration, or handling of tritium, and equipment as follows:
> (a) Hydrogen or helium refrigeration units capable of cooling to $-250°C$ (23 K) or less, with heat removal capacity greater than 150 watts or
> (b) Hydrogen isotope storage and purification systems using metal hydrides as the storage or purification medium.

The maximum quantity of tritium which is exempted form these guidelines in any chemical or physical form is 1500 GBq, slightly less than the former CoCom limit of 3700 GBq.

There are a few more items on the list which are related to tritium. Restricted are the main raw materials for the breeding of tritium, lithium-6 (2.10), and helium-3 (8.6), as well as neutron generators which make use of tritium (8.1).

The NSG guidelines have the same shortcomings as the CoCom regulations. Neither is legally binding and both have to be translated into national law to become effective. There are no provisions to verify the stated end-use or to detect removal or reexport to sensitive countries. Both instruments are discriminatory and in both cases only a limited number of countries are members of the respective group.

EURATOM/Canada agreement on tritium transfers and controls: Canada has become the world's largest producer of tritium and intends to sell it primarily for fusion energy research projects (see Section 1.4.2). In May 1991, EURATOM and Canada extended by an exchange of letters their agreement for cooperation in the peaceful uses of atomic energy of October 6, 1959 (EURATOM and Canada, 1991). It covers the field of fusion research and development. In this agreement provisions for tritium supplies are modeled after the extension which was made earlier for heavy water. In the meantime, heavy water had been placed under safeguards.

The parties agreed that EURATOM is authorized to establish control procedures for tritium shipments from Canada to EURATOM member states. "EURATOM shall apply to tritium items appropriate recording, accounting, and inventory procedures." EURATOM verifies the inventory at the receiving facility as long as the tritium remains there and makes sure that the tritium is neither taken without authorization nor used for purposes other than fusion research, nor re-transferred beyond the territories of EURATOM member states without prior written consent of the Government of Canada (EURATOM and Canada, 1991).

After the conclusion of the new agreement, a Joint Technical Working Group (JTWG)[43] was established to work out the modalities of tritium control and accounting methods. The JTWG shall regularly review the control procedures.

The purchases of tritium by the Tritium Facility Karlsruhe (5 g in 1993 and 16 g in 1997), the Joint European Torus at Culham (0.1 g in 1991 and 9 g in 1997), and the European Tritium Handling Laboratory at Ispra (possibly up to 100 g) are the precedents for these new multilateral control procedures.[44]

1.5 Rationale for international tritium control

There are several arguments in favor of international controls of tritium. The main goals can be summarized as follows: to prevent the horizontal proliferation of tritium to nuclear threshold countries on the one hand and to reduce the nuclear weapon states arsenals on the other. Both goals could eventually be merged so as to achieve and verify a nuclear-weapons-free world.

The instruments and the respective rules and procedures to achieve such controls vary with the respective goals, i.e., various reasons to control tritium call for one kind of instrument, and other reasons call for different provisions. Nevertheless, rationales for establishing international controls of tritium are given in the following list without systematically considering appropriate control instruments in this section.[45] Most of the following points are related to vertical and some to horizontal proliferation.

1. **Allowing further production of tritium by nuclear weapon states and especially the decision to construct a new production facility would send the wrong signal.** The ultimate goal should be a nuclear-weapons-free world. Tritium production policy could be a litmus test to find out whether this goal is envisaged or whether it is the national priority to ensure the supply of tritium in order to maintain a large and effective nuclear arsenal. In any case, tritium production for nuclear weapons purposes implies diversion of nuclear energy because tritium can only be produced in a strong neutron source.

   Plans to develop new tritium production facilities could be postponed at least for a decade. If START implementation proceeds as planned, there will be no need for any of the two weapons states involved to resume tritium production for more than 20 years. The U.S. will not need any new tritium before 2016 and probably far beyond that date. Despite the lack of concrete information, a similar date may be assumed for Russia. After that, annual tritium shortages will be comparatively small and the construction of a new production facility might be completely avoided.

2. **Significant costs could be saved if no new military facilities were constructed to produce tritium.** If eventually it turned out that no new tritium production is needed, significant amounts of money could be saved. In 1996, the U.S. budget contained $30 million for preparing a decision between

two options. In August 1998, the Congressional Budget Office estimated that the accelerator option would cost about $9.5 billion in 1999 dollars over a 40-year period. Using existing reactors would cost about $1.8 billion. At the end of 1998 the U.S. DOE decided to use existing commercial reactors to restart tritium production in 2003 (Ferguson, 1999).

3. **Verification of a cutoff agreement would be easier if tritium were included.** The same facilities can be used for the production of plutonium and tritium. Therefore, in the case of a verified cutoff agreement, any tritium production facilities would have to be safeguarded against plutonium production. All breeding targets would have to be checked for not containing natural uranium and for not being reprocessed in order to extract plutonium. This would probably require new safeguard methods. The "zero" approach where no military production reactors or reprocessing plants are allowed to operate would be easier to verify, possibly without the need for any on-site inspections.

4. **Tritium from retired warheads should not be reused in nuclear weapons.** The INF and START treaties do not include any provisions regarding nuclear materials gained from retired warheads. Nevertheless, the progress in nuclear disarmament has drawn attention to the question of what should be done with them. Although nuclear warheads have been excluded from the INF treaty, it is desirable to develop means to verify a reduction of the nuclear materials. Whereas plutonium and highly enriched uranium could be transferred to IAEA safeguarded stocks, there are no proven methods in place to verify the further whereabouts of tritium.

   In general, this argument is not even advocated by proponents of nuclear disarmament (see, e.g., Donnelly, 1989; and Taylor, 1989). They propose that tritium regained from retired warheads be used to maintain other warheads or stored until it decays.

5. **Tritium control in nuclear weapon states would make international tritium control less discriminatory.** Including tritium in a verified cutoff agreement (the "integrated cutoff") (see Section 1.8.3) would be a step towards nondiscriminatory international control of nuclear materials because tritium is already gaining increased attention on the level of horizontal nonproliferation (see Section 1.7).

6. **New technologies for the production of tritium could proliferate and be used for clandestine plutonium production.** If the development of new production technologies such as accelerator-based breeding systems (see, e.g., Crawford, 1989, January 27) was not brought to a halt, they would pose a new threat for horizontal nuclear proliferation because they could be used by proliferating countries for plutonium production.

7. **The radioactive decay of tritium could be used to enforce further nuclear disarmament.** The half-life of tritium is 12.3 years. Therefore, the suggestion has repeatedly been made to use its decay as a "forcing function" to set the pace for nuclear disarmament in the order of minus 5.5% per

year. Therefore, to include tritium in the cutoff agreement would constitute a stronger commitment by the nuclear weapons states to complete nuclear disarmament (see Section 1.8.1). This idea was discussed extensively in 1988 in the U.S. after the production reactors at the Savannah River Plant had been shut down for safety reasons (Nuclear Control Institute [NCI] and American Academy of Arts and Sciences [AAAS], 1988; Mark, 1988; and Sutcliff, 1988). The general conclusion of this discussion was that disarmament could not be enforced by a technical fix but would rather follow the military and political assessment of possibilities for further disarmament.

8. **The total yield of nuclear weapon arsenals could be reduced significantly by removing tritium completely.** One could even go beyond the tritium cutoff and in addition remove all tritium from the remaining nuclear weapons in order to reduce their explosive yields. Even with START II implemented, the total maximum yield of the U.S. nuclear arsenal would still be in the order of 1400 Mt, i.e., it would decrease only by a third as compared to the current stockpile (end of 1997). The removal of tritium would immediately reduce the total yield by two orders of magnitude down to 20 Mt and to $\simeq$13 Mt after START II implementation (see Section 1.6.3).

9. **Bi- or trilateral agreements between supplier and receiver countries combined with supervision by an international agency could help to stem the horizontal proliferation of tritium.** The export of tritium could be subject to safeguards according to a bilateral (supplier and receiver country) or trilateral (involving a third party, e.g., the IAEA) agreement. In the latter case, the agreement would be negotiated between supplier and receiver countries under the auspices of the third party which monitors and enforces the agreement.

   A precedent for this kind of international tritium control has already been set. The recent agreement between Canada and EURATOM gave to this European agency the mandate to control tritium shipments from Canada to EURATOM member states and to verify the end-use of the tritium in civilian fusion research programs.[46]

10. **Horizontal nonproliferation could be strengthened by introducing international tritium controls.** Tritium becomes of growing significance for horizontal nuclear proliferation. The proliferation of technology related to nuclear weapons covers a wide spectrum. The development of more sophisticated weapons including those utilizing tritium by nuclear threshold countries is gaining in significance. A country which has access to only small quantities of plutonium or highly enriched uranium could multiply the yield by boosting the nuclear explosion by fusing tritium and deuterium.

    There are several indications that nuclear threshold countries got involved in tritium technology in connection with their nuclear weapons program (see Section 1.7).

    Expanding international controls over nuclear materials to cover tritium is being considered and has partly already been started.[47] A stepwise approach could be taken to international tritium control which begins simultaneously with agreements between individual countries on both the horizontal (e.g., EURATOM/Canada) and the vertical (e.g., U.S./Russia) level, and becomes broader when more countries join the respective agreements. Eventually it might merge into a nondiscriminatory agreement to ban the production of fissionable and fusionable materials.

11. **Excess quantities of tritium from civilian sources raise the question of how nonproliferation could be ensured.** Since 1990, Canadian tritium supplies from detritiation of heavy water far exceed civilian demand (see Figure 1.2). This raises concerns about the horizontal proliferation of tritium.

12. **In the far future, fusion research may use large amounts of tritium raising the question of how nonproliferation can be ensured.** In the far future, tritium demand for peaceful uses might rise significantly if the nuclear fusion technology programs were successful. The growing market for tritium would make it increasingly difficult to separate the military and civilian uses of tritium.

    One of the largest tritium research facilities, the Tritium Systems Test Assembly (TSTA) at Los Alamos National Laboratory, has an inventory of more than 100 grams. Japan has a tritium laboratory of similar dimensions and in Europe facilities at Ispra and Karlsruhe are receiving increasing amounts of tritium. The first fusion experiments that made use of some grams of tritium were conducted by TFTR at Princeton in 1993 and by JET at Culham in 1997.

    If ever a fusion power reactor burning deuterium and tritium comes into operation, it would contain a steady inventory of a few kg and burn about 180 kg per $GW_e y$.

1.6 Reversing vertical proliferation by tritium control

1.6.1 Tritium control and qualitative disarmament

There are various proposals for qualitative disarmament under discussion. These include dealerting,[48] a No-First-Use Treaty, and withdrawal of nuclear weapons from foreign territory. Qualitative disarmament leads to marginalization of nuclear weapons. Eventually it will be a small step towards complete elimination of the remaining and qualitatively disarmed arsenal. This concept is different from quantitative disarmament, since it would not reduce the number of nuclear weapons. However, the numbers can be reduced in parallel by other agreements.

A linear scale of technical steps in qualitative disarmament is defined here by the increased time required to prepare a qualitatively disarmed nuclear weapon for delivery to a military target. In effect, a low grade of qualitative disarmament prevents the effective capability for launch-on-warning. A high grade of qualitative disarmament may eventually result in disassembling all nuclear warheads and abandoning the capability of rapid redeployment. The following steps of qualitative disarmament, with increasing time required for making the warheads ready for delivery, could be taken:

1. Technically, the first step in qualitative disarmament of nuclear weapons is to take them off alert. This can be achieved by deleting the target coordinates from the executing computer programs which control the delivery systems. This is a narrow notion of dealerting. This step can be easily and possibly even automatically reversed within minutes.

2. Dealerting can be improved by removing nuclear warheads from the delivery systems. This step can be reversed within hours.

3. The third grade of qualitative disarmament is achieved by removing the nuclear warheads from the deployment site of the delivery system. This proposal is also known as sequestration. According to the definition given above, sequestration is more effective the further away the storage site is located. Provided that the distance is large enough, the weapons can be put back on alert within days.

4. As a fourth step, nuclear warheads can be dismantled into their main components, which are stored at different places. For example, three different storage sites can be selected for the fissionable core element (e.g., the plutonium pit), the tritium ampoule, and the metal casing. All sites can be different from the place where the equipment for assembling the weapon is kept. Reversing this step will probably take weeks.

5. Yet another improvement in qualitative disarmament can be achieved by removing tritium from nuclear warheads. Assuming that all nuclear weapons rely on tritium for boosting their yield, they are rendered militarily dysfunctional and have a reduced catastrophic potential as soon as the tritium is removed. All weapons can be restored to full capability by replacing the tritium. Two variants of tritium control are possible.

    (a) If the production of tritium is banned, this kind of qualitative disarmament will affect only a limited number of warheads once the decay of the military tritium inventory has decreased below the demand of the current active arsenal. This approach is discussed in the context of a fissile material treaty in Section 1.8.

    (b) The complete arsenal could be involved if the tritium ampoules were removed from all nuclear warheads and their positions were sealed and inspected to verify that no tritium was replaced. It takes a few months to reproduce tritium and it may take years to construct a production facility in the case that no reactor is kept in cold stand-by for this purpose. The radical approach of tritium elimination is the topic of Section 1.6.3.

1.6.2 The relation of tritium and weight to yield of nuclear weapons

The consequences of a shortage or even the complete elimination of tritium from military stocks are discussed in this section. The total yield of nuclear arsenals serves as the measure for this assessment. A more adequate parameter to evaluate the impact of tritium on the military potential of nuclear weapons would be the "kill-factor," which takes into account the accuracy of delivery systems. Boosting with tritium increases the yield of the warhead while the weight can be kept low. Low weight allows a high accuracy in targeting and thus improves the kill-factor further. Therefore, the weight of warheads will be considered in the following assessment as well.

The case of the U.S.[49]

The impact of the complete elimination of tritium on nuclear weapon arsenals becomes apparent when looking at a graph which shows the yield of nuclear weapons vs. their weight (see Figure 1.4). The data used are best estimates publicly known.[50] When a range of data was given, the lower limit for the weight and the upper value for the yield were used. Thus, the highest practically achieved yield-to-weight ratios are shown in the graph.[51] The graph considers only U.S. nuclear warheads because estimates from other nations' stockpiles are not available in comparable detail. Some historic nuclear explosions are included for comparison.

In Figure 1.4, different symbols are used to distinguish nuclear weapons that make use of tritium from those that do not. There remains a slight uncertainty about the use of tritium in some thermonuclear bombs (B53, W78, W56). The three weapon types (fission, boosted[52] fission, and thermonuclear) appear in clusters. The line for a constant yield-to-weight ratio of 0.1 kt/kg seems to be roughly a dividing line between fission weapons on the lower side and thermonuclear weapons on the upper side. But this does not mark the theoretical limit of the yield-to-weight ratio for fission weapons. Assuming that 100% of the nuclear material is burned, the theoretical limit lies at 17.5 kt/kg for uranium-235 and 20 kt/kg for plutonium-239. The corresponding value for pure fusion reactions is 80 kt/kg. These theoretical limits cannot be reached in reality because additional weight is required for the chemical explosive, the bomb casing, and other parts of the weapon, and because 100% efficiency in burning the nuclear fuel cannot be achieved.

Some nuclear tests with high yields have been conducted without using tritium. The highest yield all-fission test carried out by the U.S. was the "King Shot" (500 kt).[53] Furthermore, the first thermonuclear weapons built did not contain tritium. That proves that in principle tritium is not necessary to ignite a fusion reaction. However, with modern boosted triggers, tritium may be needed in two-stage warheads.

Constraints on weight and volume posed by the delivery systems demand that the nuclear weapons be lightweight. In the early 1950s, this was achieved primarily by improving the effectiveness and reducing the required volume of the chemical explosive. In the mid-1950s, another significant weight reduction of fusion weapons was achieved by replacing their 1.5-meter-diameter unboosted primaries

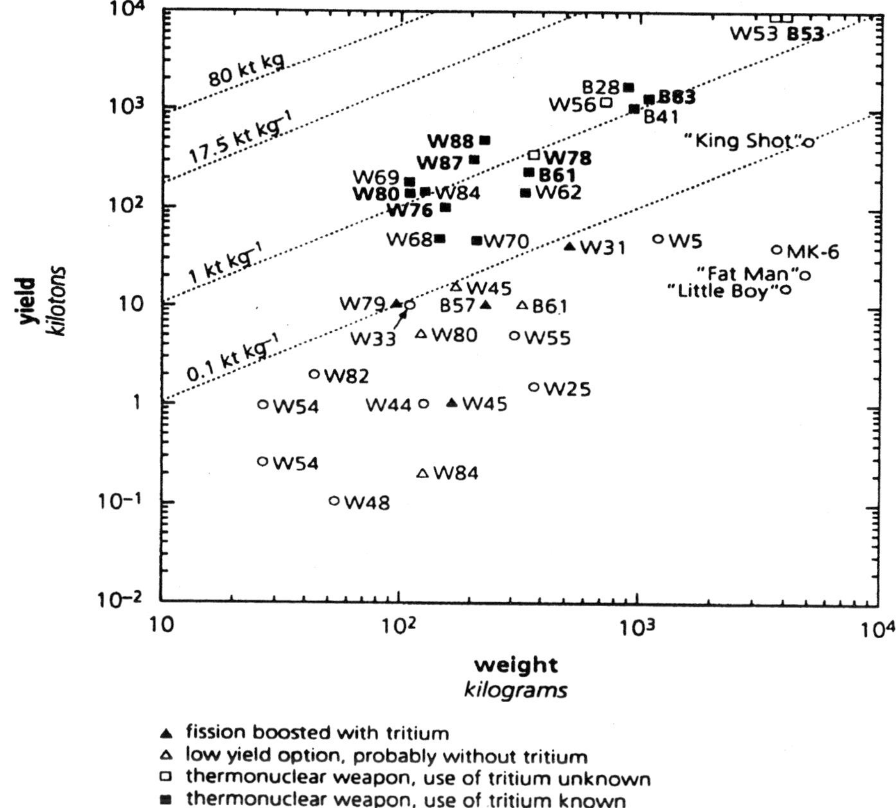

Figure 1.4 Yield and weight of U.S. nuclear warheads and some nuclear explosions. The names of warheads which were active in the U.S. nuclear stockpile in 1996 are shown in bold face. At the end of 1996, the replacement of the bomb B53 by the new B61-11 (modification 11) with a yield range from 0.3 to 340 kT was started. The retired warheads that are included in the figure have not yet been completely dismantled. Some are kept in "inactive reserve" without plans for dismantlement. Although the details are classified, it can be assumed that about 400 W84 warheads are still in inactive reserve as well as W62, W68, W69, W76, and W78 warheads. The nuclear weapons used against Hiroshima and Nagasaki as well as the nuclear test explosion "King Shot" are included for reference. The hollow symbols symbolize warheads not using tritium; the filled symbols indicate tritium use. Sources: Cochran *et al.* (1984); Hansen (1988); Norris (1993); and Norris and Arkin (1993). Figure reprinted from Kalinowski and Colschen (1995).

with smaller diameter (55 to 75 cm) boosted primaries. It can be assumed that thermonuclear weapons in the current U.S. stockpile are triggered by a boosted primary.

There are a number of ways to vary the yield of a particular nuclear weapon design, one of which is boosting. It can increase the yield of pure fission weapons or the fission primary of thermonuclear weapons — depending on the fission efficiency without boosting — by a "boosting factor" of 2 to 10.[54] For some weapon types, more than one yield is shown in the diagram (W45, W54, B61, W80, and W84). The lower yield might be due to a reduction of the amount of tritium and deuterium injected into the core of the nuclear explosion or to the total removal of the tritium ampoule. However, these values might also be achieved by the timing or intensity of the external neutron source which initiates the chain reaction. In some weapon designs (especially in older ones), it is possible to remove the nuclear core and insert a differently sized core to provide different yield options.

Therefore, there is no simple rule to estimate the reduction in maximum yield of boosted weapons when tritium is eliminated. Even if the boosting factor is known, the reduction factor is not known because it may depend on more than tritium removal. Operation with and without tritium may be optimum at different configurations, especially regarding the geometry of the pit and timing sequences of the explosion. When estimating the effect of tritium on the yield, it should be noted that the yield of a boosted primary or boosted fission weapon varies with storage time due to the radioactive decay of tritium and the simultaneous increase of its decay product, helium-3, which is a strong neutron absorber. However, this probably does not significantly affect the yield of the secondary stage in thermonuclear weapons as long as the primary is strong enough to ignite the fusion reaction. The initial tritium content is always high enough to allow for a few years' tritium decay.

The crucial question is whether the primary without tritium would still yield enough energy to trigger the fusion stage. If not, a thermonuclear device would yield no more than an unboosted fission primary — around 1 kt or even as low as 0.4 kilotons in the case of miniaturized primaries (see, e.g., von Hippel *et al.*, 1987). This is highly probable and it is therefore safe to assume that the yield of thermonuclear weapons, which on the average is about 300 kt, could decrease by a factor of 100 when tritium is eliminated and the design is not changed to compensate.

The case of other nuclear weapons states

The impact of a tritium shortage or tritium removal on the nuclear arsenal of the U.S. is further discussed in the following subsection. It remains to be seen how tritium elimination would affect the nuclear stockpiles of the other nuclear weapons states. The high yields of the strategic warheads of Russia, ranging from 100 to 750 kt, indicate that they are all two-stage thermonuclear warheads. It is very likely that they all depend on tritium, i.e., the thermonuclear stage is triggered by a tritium-boosted fission primary. On average, the more than 6000 strategic nuclear warheads remaining in the active arsenal of Russia by the end of 1997 have a yield of 430 kt (Norris and Arkin, 1998). Without sufficient tritium remaining (a quantity which is not publicly known), the second stage would probably not ignite and the yield of the thermonuclear warheads would basically be reduced to the unboosted yield of the fission primary. If this is around 1 kt, the yield reduction due to the lack of tritium would exceed a factor of 100. It would possibly be higher than in the case of the U.S. because Russian strategic warheads have on average a larger maximum yield than U.S. warheads.

1.6.3 Consequences of yield reduction by elimination of tritium

The impact of tritium elimination on a nuclear weapons arsenal is demonstrated by using the U.S. as the example.[55] As a result of the considerations described in the previous subsection, this assessment is based on the following assumptions, which seem reasonable but cannot be proven by the author:

- There are only two-stage thermonuclear warheads left in the current active arsenal, and all of them have tritium-boosted primaries.
- The primary of a thermonuclear weapon cannot yield sufficient energy to trigger the fusion stage when tritium is missing. Therefore, a yield reduction by two orders of magnitude due to the removal of tritium can be assumed. The yield of unboosted primaries is lower by a factor of 100 than the upper yield of the thermonuclear weapon and smaller than 10 kt.
- There is no substitute for tritium. One source (Mark, 1988) states that "isotopes other than tritium, such as helium-3, have been considered for boosting, but use of these is considered not to be within reach of present weapons technology."

To get an idea of the qualitative effect of tritium removal on the U.S. nuclear arsenal, see Table 1.5. The number of warheads is multiplied by a yield, and all yields are added to provide the total yield. This calculation is made for the upper and lower bounds of given yield ranges as well as for the case of tritium removal. For the latter case, a crude estimate is made by reducing the upper yield bound for thermonuclear warheads by a factor of 100, but down to at most 10 kilotons.

Total yields are presented for three cases:

1. U.S. stockpile as of end of 1997
2. projected estimate of a post-START II arsenal
3. nuclear arsenal after further deep cuts down to 500 warheads

START II levels are used for illustration. This treaty will not be implemented. Instead, the Moscow Treaty of 2002 will limit the strategic nuclear weapons to 1700–2200 in the active arsenal by the year 2012. However, no single warhead is required to be destroyed. According to these estimates, even after START II, the total yield (upper bound) would be decreased only by a third by 2003, whereas the removal of tritium would reduce the total yield by two orders of magnitude at once. Even a stockpile of 500 warheads with tritium would still have a total yield which is one order of magnitude higher than the stockpile as of end of 1997 without tritium (see Table 1.5).

The military usefulness of a particular nuclear warhead without tritium is reduced even more than its yield. The military mission assigned to a warhead may not be achievable when the kill-factor, i.e., the product of yield and targeting accuracy, is too low to reach the desired goal, e.g., to destroy a hardened missile silo with a certain probability. In this sense, nuclear weapons without tritium are dysfunctional

Table 1.5 Total yield of operational U.S. strategic and tactical nuclear weapons stockpile under different assumptions (numbers in brackets refer to a projection for a hypothetical START II implementation at the end of 2003).

warhead type	weapon system	number in stockpile end of 1997[a] (after START II)[c]	nominal yield [kt] range[b]	total yield [10^3 kt] upper bound	total yield [10^3 kt] lower bound	total yield [10^3 kt] without tritium, rough estimate
B61-7/-11	strat. bomb	500 (450)	0.3-340	170 (153)	0.15 (0.14)	1.7 (1.5)
B61	tact. bomb	600 (100)	0.3-175	105 (18)	0.18 (0.03)	1.1 (0.2)
W62	Minuteman III	600 (0)	170	102 (0)	102 (0)	1.0 (0)
W76	Trident I C4, II D5	3,072 (1,280)	100	307 (128)	307 (128)	3.1 (1.3)
W78	Minuteman III	900 (0)	335	302 (0)	302 (0)	3.0 (0)
W80	ACM/ALCM	800 (350)	5 and 150	120 (53)	4 (2)	1.2 (0.5)
W80	SLCM	350 (400)	5 and 150	53 (60)	2 (2)	0.5 (0.6)
B83	strat. bomb	500 (500)	low-1,200	600 (600)	5 (5)	5.0 (5.0)
W87	MX/Peacekeeper	500 (0)	300	150 (0)	150 (0)	1.5 (0)
W87	Minuteman III	0 (500)	300	0 (150)	0 (150)	0 (1.5)
W88	Trident II D5	384 (400)	475	182 (190)	182 (190)	1.8 (1.9)
(1) end of 1997		8,206		2,091	1,054	19.9
(2) at START II level		$\simeq$4,000		$\simeq$1,350	$\simeq$480	$\simeq$13
(3) after Moscow Treaty 2012		$\simeq$2,500		$\simeq$600	$\simeq$290	$\simeq$6
(4) after deep cuts to 500 warheads		$\simeq$500		$\simeq$150	$\simeq$50	$\simeq$1.5

[a] Numbers are taken from Norris and Arkin (1994).

[b] Figures are taken from Norris and Arkin (1994) with a change related to B61.

[c] Numbers in brackets are projections for after START II implementation as estimated by Norris and Arkin (1995).

with respect to their dedicated military mission. However, their destructive potential remains unimaginably devastating. They are still weapons of mass destruction, and the military could think of many ways of using such a weapon.

Consequently, after tritium production at the Savannah River Plant was stopped in 1988, U.S. government officials considered this a national emergency. The urgency disappeared when it became apparent that there is enough tritium in the production pipeline and released from nuclear disarmament for at least the following two decades.

On the assumption that leftover tritium would be redistributed to serve the maximum number of warheads possible and making further the over-conservative assumption that warheads without tritium would effectively be useless, the deployment of nuclear warheads would decline at the rate of tritium decay, i.e., by 5.5% per year.

1.6.4 Yield reduction by tritium elimination: possibilities for qualitative nuclear disarmament

Complete elimination of tritium from nuclear weapons arsenals[56]

Though a production ban on fresh tritium supplies may also have disarmament effects, it is dealt with in the context of a fissile material cutoff treaty in Section 1.8. The reduction of the yield of nuclear weapons by the elimination of tritium could be used for a novel approach to nuclear disarmament. The precondition for such a qualitative nuclear disarmament is the decision to abandon high yield nuclear weapons and to reduce the overall yield of the arsenal significantly.

This approach to nuclear disarmament was presented by Russian nuclear weapons experts in 1991 (Trutnev *et al.*, 1991). Instead of reducing the number of delivery systems or nuclear warheads, Trutnev *et al.* suggest limiting the explosive yield of each nuclear weapon to three to five kilotons TNT. Theoretically, the total explosive power of the nuclear arsenals of the U.S. and Russia — about 5 gigatons total yield as of end of 1997 — could be reduced by a factor of 100. They argue that such nuclear arsenals would still be adequate to sustain the doctrine of deterrence, but they would no longer pose a threat to civilization. Furthermore, the balance of power would be more stable because the potential first-strike effectiveness of strategic offensive forces would be substantially reduced. Trutnev *et al.* believe their proposal would be a decisive step towards a world free of nuclear weapons. The next step would be the numerical limitation of nuclear warheads, while preserving bilateral stability. Eventually the military concepts and political doctrines that rely on nuclear deterrence must be abolished to clear the way for the elimination of nuclear weapons.

The present strategic U.S. nuclear arsenal has a total maximum yield of about 2.1 gigatons (see Table 1.5). If used in a strategic war, the total yield of the U.S. and Russia — as of end of 1997 — is about 5 gigatons and could lead to a "nuclear winter." All models of a strategic nuclear war which would lead to a nuclear winter scenario assume total explosive yields in the range of 5 to 10 gigatons, which would kill 750 million to 1.1 billion people in the northern hemisphere immediately and probably another two billion later (Ehrlich, 1983). Even with START II levels, the active nuclear arsenal remaining in the U.S. would probably still have an upper bound of total yield on the order of 1.4 gigatons and 0.6 gigatons after implementing the Moscow Treaty by 2012 (see Table 1.5). This reduced yield probably does not preclude the possibility of a nuclear winter.

Some warheads remaining in the stockpile have selectable yields. If, after reaching START II levels, yield selections above their lowest options are disabled, the total U.S. yield would be reduced to 0.5 gigatons and further down to 0.3 gigatons with implementation of the Moscow Treaty (see Table 1.5). It is not clear if this could be implemented in an irreversible or verifiable way, and this would bring the world only a modest step away from the horrific scenario of a strategic nuclear war.

Obviously, the feasibility and verifiability of reducing explosive yields are the crucial problems for any approach to disarmament. Trutnev *et al.* (1991) claim that their proposed yield limit could easily be implemented making use of the physical construction peculiarities of the modern nuclear weapons developed in the former U.S.S.R. and the U.S. Although the authors do not provide the details, readers can trust in their work because they have access to classified information. In fact, Trutnev is known as the "designer of the Soviet thermonuclear arsenal."

The assertion of this monograph is that **tritium elimination might be the key to both feasibility and verification of a yield reduction** as a qualitative nuclear disarmament measure.[57] The best way of achieving and verifying a substantial reduction of the explosive yield might be the *complete* elimination of tritium from all nuclear weapons including those already withdrawn, as well as from the nuclear weapons production cycle.[58] The yield would immediately decrease by a factor of 100 and be 0.020 gigatons with the current stockpile, 0.013 gigatons after reducing down to START II levels, 0.006 gigatons after implementing the Moscow Treaty, and 0.0015 gigatons with 500 warheads (see Table 1.5).

Such qualitative disarmament would be of high value even if tritium is kept more or less ready to reinsert, because high-yield nuclear weapons could not be used quickly and a further step for possible nuclear escalation is introduced.

The idea of using tritium decay as a forcing function was discussed extensively in 1988 (see Section 1.8.2). But the basic idea of the "forcing function" is different from this proposal in that it was intended to force a reduction in numbers of warheads to keep pace with the radioactive decay of tritium, whereas this proposal is independent of the actual number of warheads that remain in the stockpile. It reduces the military usefulness and the catastrophic potential of the whole nuclear weapons arsenal and is reversible by reinserting and, if necessary, reproducing tritium. Furthermore, the decay takes years, whereas elimination achieves immediate disarmament.

Implementation and verification of complete tritium elimination

Some problems of the tritium approach to reduce the yield are obvious. It is difficult to estimate the potential of tritium-deprived nuclear arsenals, especially because the technical assumptions made in this book may not hold or cannot be proven because the relevant information is kept classified. This is necessary to facilitate nonproliferation and should not be altered. It is not even realistic that the U.S. and Russia would want to share weapon design information. Tritium elimination would not be easy to implement as a bilateral verified agreement, because not only may Russian and U.S. warheads be affected in an asymmetric way,[59] but it may be difficult to negotiate a verification procedure that includes on-site inspection of warheads.

If a single country decided to do without high-yield weapons, it could eliminate tritium unilaterally in a reversible manner, and there would be no need for verification. This could encourage other countries to do the same. The advantage of this approach is that it does not depend on an elaborate, verified agreement like the Integrated Cutoff (ICO) (see Section 1.8.3), which focuses on production controls, but rather is based on unilateral, possibly reciprocated decisions to take tritium out of nuclear arsenals. This can be done stepwise, e.g., by first declaring a production moratorium, and then removing tritium from withdrawn warheads and the production cycle, and eventually from all deployed warheads. Some transparency regarding tritium quantities involved and aggregated data on stockpile effects will certainly help in building confidence, but weapon design information can still be kept secret.

If access to nuclear warheads were permitted to inspectors, tritium elimination could be verified. The removal of the tritium is a simple process because tritium ampoules are designed to be replaced. The tritium can be kept ready to reinsert but moved to a central verified storage. The positions of tritium ampoules in the warheads can then be sealed, and the absence of tritium can be verified by a test of seal integrity. Inspections of warheads would not reveal information on their internal design.

Remaining tritium stocks could be handed over to an international inspection agency. Since tritium decays with a half-life of 12.3 years, undeclared stockpiles would be depleted and eventually fall below a significant amount. Complete verification would have to detect any clandestine production or transfer of tritium from civilian stocks for military purposes. Tritium can be integrated in a verified cutoff for the production of fissile materials (see Section 1.8.3). Military production reactors could be shut down and verified by national technical means as well as on-site inspections. Chapter 3 shows that tritium control procedures could be established which are able to detect quickly any significant diversion of tritium.

Conclusions on complete tritium elimination

Some problems of the tritium approach to reduce the yield of nuclear weapons are obvious:

1. The technical assumptions made here may not hold or cannot be proven because the relevant information is kept classified.

2. It may be difficult to estimate the potential of the remaining stockpiles.

3. It would not be easy to implement this proposal as a bilateral and verified treaty. Not only may Russian and U.S. warheads be affected in an asymmetric way, but it may also be difficult to negotiate a verification procedure including inspection of warheads.

4. Break-out from such an agreement would always be possible if any country decides to do so. As long as nuclear reactors are available, a dedicated effort could be undertaken to produce strategically significant quantities of tritium.

On the other hand, there are a number of advantages which would make a yield reduction by tritium elimination an attractive way to complete nuclear disarmament:

1. The world would be brought a substantive step away from the otherwise continued danger of a nuclear winter and the other dramatic effects caused by an accidental or even deliberate strategic war.
2. The mechanism can be applied right away without long preparations. Exchangable tritium capsules simply have to be removed from the warheads.
3. Verification of tritium nondiversion with satisfactory detection capabilities can be achieved for the case of a bilateral or international treaty, and the trustworthiness of actions can well be demonstrated in the case of a unilateral step.
4. Money could be saved by stopping the further development and operation of new production facilities for tritium. Also, the proliferation risk in developing new production technologies (e.g., accelerators), which might then spread to other countries and be used for plutonium production, can be avoided.
5. Tritium elimination may serve as the first step for the introduction of an international tritium control system (ITCS) (see Section 1.7.2 and Colschen, 1995). With such an instrument the de facto nuclear weapons states might be hindered effectively from progressing towards boosted nuclear weapons and thereby increasing the yield of their stockpile, which is mostly limited by the small amount of plutonium and highly enriched uranium available to them.

The imminent advantages of this nuclear disarmament approach make it worthwhile to go into a more elaborate assessment of the pros and cons.

Eliminating tritium from specific warheads

There are various possibilities for reducing the yield of particular types of warheads. The elimination of tritium could serve this purpose as well. A yield reduction that does not apply to the whole nuclear arsenal could make sense. For example, it was suggested as part of a strategy to put nuclear weapons off hair-trigger alert (Blair *et al.*, 1997). Blair *et al.* suggest, as one of several points, replacing the W88, which has a yield of 475 kt of TNT, by another warhead that has a lower yield. For example, the W76, which has an upper yield of 100 kt, could be used. The reason for this suggestion is that the U.S. warheads that most threaten Russia's nuclear deterrent are the nearly 400 high-yield W88 warheads which are deployed on the Trident submarines (see Table 1.5). The W88 is the weapon with the second largest yield in the U.S. nuclear arsenal. In the current arsenal it accounts for 10% of the upper yield of the whole arsenal. The B83 bomb is even worse; it contributes nearly a third to the total yield.

The yield reduction could be achieved by replacing the warhead by another type with a lower yield. The verification of the replacement of the W88 by another type could be done on the basis of the smaller size of the W76 (von Hippel, 1998). The weight difference between the W76 and the W88 might also be used to verify the suggested replacement. For comparison, the high-yield strategic bomb B83 has a much higher weight. If a weight limit is set anywhere between 400 and 1000 kg, this would imply the withdrawal of the high-yield strategic bomb B83 and could reduce the total upper yield of the U.S. arsenal by a third. Several warheads have weights between 100 and 400 kg. Therefore, a weight limitation of 100 kg would leave nothing in the stockpile, because there are no warheads left in the stockpile below this weight.[60]

Another way of achieving the yield reduction in a verifiable way could be the elimination of the tritium ampoules. The resulting unboosted fission weapon has too low a yield to trigger the fusion stage. The yield would go down by two orders of magnitude. Since this is easily reversible, the navy may be prepared to undertake such a step. If verification is wanted as a confidence-building measure, it is possible to seal the location of the exchangeable tritium ampoules and permit inspectors to check the integrity of the seals.

The fact that the yield reduction of particular types of warheads is suggested in the context of taking weapons off hair-trigger alert provides further evidence that it makes sense to take the view that tritium control is related to dealerting as a way of qualitative disarmament. The time span of this yield limitation as a kind of dealerting is dependent on the time required to replenish the tritium reservoirs of the warheads.

1.7 Horizontal nonproliferation of tritium

1.7.1 Tritium-related activities in de facto nuclear weapons states

De facto nuclear threshold countries which have achieved and utilized their technical capabilities to manufacture a simple fission weapon and which integrate tritium into their nuclear weapons programs undergo vertical proliferation towards second-generation nuclear weapons which make use of fusion energy. Such a step requires significantly more physical and technical sophistication, and some analysts suspect that nuclear weapons which make use of tritium would not work reliably without nuclear testing.[61] The fact that several de facto nuclear weapons states are known to engage in tritium technology within their weapons programs without nuclear testing seems to falsify this assumption. There are indications that most countries which are suspected of having developed nuclear weapon capabilities have also engaged in acquiring tritium and tritium technology to enhance these capabilities. India and Israel are indigenously producing tritium. Iraq undertook the first steps necessary for an indigenous production of tritium. Pakistan and South Africa imported tritium (see below). Obviously, tritium is gaining significance for vertical proliferation.

All five recognized nuclear weapons states succeeded first in the construction of a boosted weapon before managing a thermonuclear explosion. The second step took less time than the first step for all of them.[62] Thus, the acquisition of tritium for suspected use in a nuclear weapons program can indicate that a country is close to managing thermonuclear explosions.

There are several technical and military arguments in favor of making use of tritium in small nuclear arsenals.[63]

- A country with limited access to plutonium and highly enriched uranium could multiply the efficiency (and thus the yield) of these materials by boosting the nuclear explosion with the help of neutrons resulting from the fusion of tritium and deuterium (see Section 1.3.2). Less fissionable material is required to achieve the desired nuclear yield.

- Boosted nuclear weapons have a smaller core compared to unboosted fission weapons with the same yield. The smaller core can be compressed with the surrounding chemical explosives in a more reliable way.

- Boosted nuclear weapons are smaller and lighter and can thus be used on missiles and artillery shells as delivery systems without compromising the yield.

- Tritium capsules can be either inserted or not, thus allowing selection of the yield. By doing this, "surgical" weapons can be tailored to the military requirements.

India may be using its research reactors Dhruva and Cirus to breed tritium from lithium-6,[64] but it is known to remove tritium from heavy water which was tritiated during normal operation in India's CANDU-type power reactors. The detritiation plant is believed to be a small-scale pilot plant. It was under construction in 1989 (Albright and Zamora, 1989). If all Indian reactors ran at a capacity factor of 70%, some 150–230 g tritium would be produced in their heavy water per year.

Israel could produce tritium at any rate between 0 and 500 g/y in its research reactor at Dimona alternatively to the production of plutonium. From the disclosures made by M. Vanunu in 1986 it is known that Israel enriches lithium to breed tritium from lithium-6 targets (revealed, 1986, October 5 and October 12).

Tritium production for nuclear weapons does not appear to be a problem for Israel, since it secretly exported some 30 g of tritium to South Africa during 1977 and 1978 in exchange for 500 to 600 tons of yellowcake (PPNN Newsbrief, first quarter 1994, p. 12).

Iraq undertook research in lithium-6 enrichment. The facility was able to process approximately 0.5 to 1 kg of natural lithium per year.[65] But the Iraqi nuclear weapons program was stalled during the second Gulf War and almost completely eliminated by the subsequent UNSCOM inspections and still was under international supervision until the inspection team left Iraq during an escalating conflict in December 1998. Suspicions that Iraq may have resumed its nuclear weapons program were ended by the occupation with coalition forces in spring 2003.

Pakistan illegally received exported tritium and tritium-handling facilities from West Germany, as became public in December 1988 (see, e.g., Hibbs, 1989a, 1989b; and Deutscher Bundestag, 1990). Between 1985 and 1987, the company Neue Technologien GmbH illegally exported 0.8 g of tritium as well as some tritium technology to Pakistan, which requested a total of 100 g. Also in 1986, Pakistan reportedly received tritium from China.[66]

Pakistan apparently also tried to acquire the capability for indigenous tritium production. It ordered from the Swiss company Sulzer, the world's only supplier of detritiation facilities, a facility to remove tritium from the heavy water of its

KANUPP reactor. This deal did not occur, presumably because the costs were too high ($30 million) (Hibbs, 1989).

If the KANUPP reactor runs at a capacity factor of 70%, some 15–21 g of tritium are produced in its heavy water per year. From the start of its operation in 1971 until September 1989, this reactor produced no more than 5126 GW_eh (Müller and Hossner, 1990) and a maximum of 150 g of tritium (not decay-corrected).

South Africa's armament corporation Armscor built six primitive, tritiumless "gun-type" nuclear devices during the 1980s. There are several strong indications that South Africa also intended to proliferate vertically. South Africa had pilot projects for the production of tritium and lithium-6 (PPNN Newsbrief, first quarter 1994, p. 12). It undertook theoretical studies to boost the yield of gun-type weapons from less than 18 to roughly 100 kilotons, and it admittedly received some 30 g of tritium from Israel (see above) as well as more tritium from unspecified "overseas suppliers" (Albright, 1994). Apparently, the tritium of Israeli origin was never used by Armscor within the weapons program and with about a third of the imported tritium lost to natural decay, the Atomic Energy Commission (AEC) of South Africa decided to convert the remainder to peaceful purposes, namely radio-luminescent safety signs, in the mid-1980s (Albright, 1994). In anticipation of the loss of political power, the white South African government officially cancelled its nuclear weapons program in late 1989 and joined the NPT in 1991.

1.7.2 International tritium control for nonproliferation

The universal approach

The situation of horizontal proliferation described in the previous section in combination with the rationale given in Section 1.5 calls for improved international coordination to ensure nonproliferation of tritium for nuclear weapon purposes. For this purpose, an International Tritium Control System (ITCS) has been proposed.[67] The *goal* of the ITCS is to detect and deter illegal diversions of tritium from civilian facilities for military purposes.

The four *rules* of this tritium control system are

1. No tritium produced in civilian facilities will be made available for any nuclear explosion purposes anywhere.[68]

2. No tritium will be exported to states not party to the treaty.[69]

3. States party to the treaty may acquire tritium by import or indigenous production for civilian purposes, provided they carry out accountancy measures, report the data (including technical data of the state of the facilities, declaration of production capacities, and actual production, important especially for heavy water reactors and related extraction facilities, declaration of present tritium stocks, accountancy records) to the supervising international agency, and accept inspections of all their tritium facilities and stocks.

4. If the accumulated amount or throughput of tritium (including imports and indigenous production) in a state party to the treaty exceeds the significant quantity (SQ, e.g., of one gram), the tritium will be subject to inspection. This includes the verification of the end-use of the exported tritium.

For nuclear weapons states party to the ITCS, these obligations apply only to the civilian facilities and materials. Under the ITCS nuclear weapon uses of tritium originating from military production, facilities in nuclear weapons states remain uncontrolled. They are addressed by the Integrated Cutoff (ICO) (see Section 1.8.3). Thus, ITCS and ICO are complementary to each other. Complete and nondiscriminatory international control of tritium would be achieved by a combination of these two agreements.

Since the verification tasks of the ITCS are comparable to those carried out by the IAEA for plutonium and HEU, as indicted above, it seems worthwhile considering such a "tritium-mandate" for the IAEA. The IAEA is the main international organization within the nuclear nonproliferation regime that carries out verification tasks worldwide and is not restricted to a specific region, such as EURATOM. In fact, if the NPT was redefined to cover tritium, the IAEA would automatically become the responsible verification agency.

In Art. III.1 of the NPT it is stated that safeguards should be applied on all source or special fissionable material[70] which do not include tritium. However, according to the same article of the NPT, the purpose of the treaty is to prevent the "diversion of nuclear energy from peaceful uses to nuclear weapons or other nuclear explosive devices." This could be interpreted such that nuclear energy may not be used to produce tritium for nuclear weapons purposes by nonnuclear-weapons states party to the treaty.

Outside the NPT it is less problematic for the IAEA to get a mandate to control tritium. The verification tasks regarding tritium are compatible with the principles and norms of the IAEA Statute. According to Article III.5 of the IAEA Statute, the Agency is authorized "to establish and administer safeguards designed to ensure that special fissionable and *other materials*, services, equipment, facilities, and information made available by the Agency (...) are not used in such a way as to further any military purpose" (emphasis added). Tritium controls could be implemented at the request of the parties to any bilateral or multilateral arrangement in the field of atomic energy (IAEA Statute, Art. III.A.5).

However, the inclusion of tritium would necessitate an amendment of the model agreement INFCIRC/153 from 1971, or a new model agreement could be drawn up as an additional protocol to the NPT and signed by the states party to the ITCS and the IAEA. The latter could be done by using the structure of INFCIRC/153 and applying it specifically to tritium. Then, all rights for the IAEA as the verifying agency for plutonium and HEU would apply to tritium likewise, including the right for "special inspections" (INFCIRC/153, paragraph 73).

If the IAEA was not the verifying agency, a new instrument would have to be created to perform these tasks. But the creation of a new international agency with a "tritium control mandate" would be difficult in terms of costs and political acceptability. Since there does not seem to be high political acceptance for a separate "tritium treaty" or a verification agency solely for tritium verification purposes, the IAEA seems to be the "natural" solution and should therefore be considered to be the verifying agency.

The required verification measures as derived from a comprehensive diversion path analysis (see Chapter 2) entail the verification of nonproduction and of nonremoval from existing and declared inventories of tritium in civilian facilities (see Chapter 3). For the horizontal nonproliferation of tritium it can be assumed that the significant quantity of tritium is in the order of the amount required for one boosted weapon, i.e., at least one gram. As a plausible estimate, a detection time of one year should be considered sufficient.

In this monograph the technical feasibility of verification is assessed and compared to the standards agreed upon for nuclear safeguards (see Chapter 3).

Regional approach: The case of Southeast Asia

After the nuclear weapons tests of India and Pakistan in May 1998, the danger of a nuclear arms race in Southeast Asia increased. A few years earlier, it had been suggested that both countries could commence nuclear arms control measures by agreeing on a ban on tritium production.[71] Both countries have acquired tritium and tritium technology for their weapons programs and have thus embarked on vertical proliferation by advancing towards boosted fission and possibly even to thermonuclear weapon designs. Tritium has strategic significance, because warheads can be built smaller and lighter while retaining the same yield (see Section 1.3.2). The largest of the Indian tests, code-named Shakti-1, was officially characterized by India as a thermonuclear test with a yield of 43 kt TNT. According to seismic recordings the maximum yield was estimated to be about 9 to 16 kt of TNT (Barker *et al.*, 1998). This is significantly lower than one would expect from a successful thermonuclear device. None of the tests of India and Pakistan was announced to be a boosted fission device.

The future of the subcontinental race can be guessed from the sequence of events which took place in the U.S.–U.S.S.R. arms race. In the initial period both sides are expected to increase their stockpiles of first-generation fission bombs, making them maximally efficient within the limits of the basic design. At the second stage, a new and more destructive type of bomb will be developed — the boosted fission weapon. This allows the construction of smaller and lighter nuclear weapons that fit in a wider variety of delivery systems. The third stage is the fusion bomb, or hydrogen bomb, of yet greater power. Subsequent stages would involve maturation of design technologies to produce bombs of exactly programmed yields, both explosive and radiation.

Many Indians and Pakistanis feel that creating nuclear arms control agreements should be abandoned in the present climate of tension and fear. They argue that this is an exercise in futility given that the pressures brought to bear upon both countries, particularly on Pakistan by the U.S., have brought about no perceptible change in their respective nuclear postures. Probably, this is unnecessarily pessimistic. But at the same time, given the level of India–Pakistan hostility, one needs to shed illusions that dramatic steps are feasible. Under these circumstances, the only way to proceed may be through cautious steps which do not call for significant changes in the India–Pakistan nuclear equation, but which nevertheless have nontrivial implications for future arms control developments. This may be as far as either country is willing to go at present, but could still be extremely significant.

Such a cautious step could address specifically the second stage of weapons development, mentioned above, because this is a point where it is relatively easy to build a sort of barrier against vertical proliferation in both India and Pakistan.

The tritium control proposal: This could be achieved by a tritium control agreement. Pakistan and India would agree that neither country will try to produce tritium, or attempt to acquire it from any source, for military purposes. Certainly, this would have no impact on the nuclear readiness of either country because the agreement would not affect first-generation weapons. Since neither India nor Pakistan is known to be seriously pursuing the design of boosted weapons after the nuclear test explosions of May 1998, it would have few immediate implications for their respective nuclear programs. The design of boosted weapons is far more complex than for pure fission devices, and actual physical testing is most likely required. But after the test series of May 1998, neither country can dare to explode a further test device and it is likely that both countries would accede to the CTBT. Therefore, in realistic terms, boosted weapons are not really a viable option for future development in Southeast Asia.

What would be gained from an India–Pakistan treaty that does not attempt to change nuclear realities and which appears so inconsequential as to appear almost irrelevant? First, it would be a breakthrough in a situation that appears to be tightly deadlocked. There is virtually no communication between India and Pakistan on nuclear matters, a highly dangerous state of affairs. Second, the agreement could act as a barrier that stops, or at least greatly slows down or postpones, the suspected development of boosted weapons and the otherwise inevitable upwards spiral. There is no question that a tritium ban would have impact on the nuclear programs of Pakistan and India, because tritium-related activities are known in both countries (see Section 1.7.1). Third, any type of nuclear agreement would be important as a confidence-building measure and would serve to make possible more meaningful future treaties when political tensions ease.

Instead of trying to achieve a tritium ban, more potent and meaningful proposals, currently being debated, to freeze fissile material production globally, and hence also regionally may be desirable. Unfortunately, though negotiations could have started at the Conference on Disarmament in Geneva in August 1998, this agreement will not easily and soon be achieved. India and Pakistan in particular will have difficulties in meeting their interests. A bilateral or regional agreement is no less improbable. While it is true that India had proposed a global cutoff of fissile materials many years ago, it has also made it clear that an appropriate cutoff would not include accounting for, much less disposal of, existing stocks of fissile material. But, as argued by Pakistan, this would freeze a situation of relative asymmetry to Pakistan's permanent disadvantage and hence be unacceptable. A bilateral fissile material cutoff agreement between India and Pakistan is much more difficult to imagine than an agreement that focuses on tritium control.

Verification: A first step in bilateral tritium control between India and Pakistan might be to expand the current control mandate of the IAEA in these two countries to cover tritium. The IAEA conducts safeguards at selected facilities in both countries according to INFCIRC/66, the model control agreement of members of the IAEA. Therefore, it is reasonable to assume that the IAEA would get the mandate to verify the nondiversion of tritium as well. Alternatively, these controls could be conducted by a bilateral verification agency.

Expanding nuclear safeguards to cover tritium control is technically a straightforward and inexpensive undertaking and should not meet any relevant political resistance. The delicate question is whether tritium control can be made complete by expanding the inspection activities to all relevant facilities without compromising national security by disclosing any information relevant to fissile material production.

Uranium enrichment facilities are not relevant for tritium control. Theoretically spent fuel reprocessing facilities could be relevant for tritium diversion, if the annual throughput of spent fuel is more than about 120 t of heavy metal with high burnup. Pakistan is believed to have done no more than experimental-scale plutonium separation. India has large reprocessing capacity.[72] When fully operating, up to 10 g of tritium generated by ternary fission could be processed each year. This amount may be enough to boost three to five nuclear weapons. Practically, it is very difficult to recover this tritium. Therefore, this production path can be neglected and inspections at reprocessing plants are not required.

Most relevant are detritiating facilities for heavy water. The required controls can be achieved by state-of-the-art accountancy procedures. At heavy water reactors, which are currently not all under safeguards in India, the tritium content of the moderator and coolant would be verified. The necessary concentration measurements can be carried out without revealing sensitive information on the operation of the reactor, especially on plutonium production.

The nonbreeding of tritium from lithium-6 can easily be verified at reactors which are already under safeguards. Routine measurement technologies such as the Neutron Coincidence Collar can reveal the relevant information (see Section 3.3). At those reactors which are not already under safeguards the nonbreeding of tritium has to be verified without revealing information about possible plutonium breeding. This control task is the most sensitive with regard to the country's attempts at keeping up an ambiguity about plutonium production for weapons purposes. The reactors concerned would be visited by inspectors for the first time. They would have no access to the reactor besides inspecting the entrance. The only instruments required are portal monitors. The detection goal would be to make sure that no significant amount of lithium-6, the raw material for tritium, enters the facility. This could be achieved by nondestructively investigating all targets for lithium. It might be possible to set up a measurement technology which could identify the presence of lithium without revealing information relevant for plutonium control. The proposed technology is based on nuclear resonance absorption (see Section 3.3), which is known to be effective for nondestructively detecting the presence of plastic explosives in luggage at airports. The development of such control procedures would close loopholes in tritium control.

However, complete verification of a tritium agreement may be a potential stumbling block. As long as a verified production cutoff for plutonium has not been agreed, tritium cannot be monitored without introducing new safeguard procedures, especially at reactors which are not currently under safeguards. The chances of agreeing to safeguards which could yield relevant information with regard to possible plutonium breeding will increase as soon as that fissile material is put under control as well.

Until that is achieved, even an unverified agreement would have its merits. A statement of intent that each country will respect its terms might be a starting point. Cheating is certainly possible but, because tritium is not a priority issue at present, this is not highly likely. Verification procedures may be invoked step by step and may, for example, commence with control of tritium generated in heavy water as a by-product. This may pave the way for the more important ban on fissile materials production.

1.8 Nondiscriminatory tritium control within a Fissile Material Treaty

1.8.1 The role of tritium within a verified production cutoff for fissile materials

The universal approach

A ban on the production and storage of nuclear-weapons-usable material is at the top of the international agenda for nuclear disarmament and nonproliferation. Historically, the main intention of a cutoff agreement was to cut the nuclear arms race at its source and to freeze the arsenals of the recognized nuclear weapons states. Since these countries now have excess stocks of nuclear-weapons-usable materials, the main effect of a cutoff agreement would be to get the nuclear programs of threshold states under international control.

For more than 30 years a production cutoff for fissile materials, namely plutonium and highly enriched uranium (HEU), has been debated with the intention of cutting the nuclear weapons production at source.[73] IAEA-type safeguard systems could be applied to verify that no country diverts sufficient amounts of fissile material to produce new nuclear weapons. In the 1960s, the U.S. made several proposals for a bilateral halt in the production of fissile materials for nuclear weapons purposes. At that time, all cutoff proposals were refused by the Soviet Union. In the 1980s, it was just the other way around. The Soviet proposals for a cutoff were refused by the U.S., beginning on June 15, 1982, when Soviet Foreign Minister A.Gromyko suggested a cessation of the production of fissile materials as a useful initial stage for a nuclear disarmament program.

Meanwhile the situation has changed (see, e.g., Albright and Paine, 1988). A real nuclear disarmament process has started. Encouraged by the progress in nuclear weapons reductions (INF, START I and II,[74] unilateral withdrawal of tactical and other nuclear weapons), further proposals for deeper reductions are under discussion and even the vision of a nuclear-weapons-free world has gained new impetus. In this context, the cutoff proposal is back on the political agenda. At the beginning of the 1990s, several bills related to a cutoff were introduced in the U.S. Congress,[75] and on September 27, 1993, President Bill Clinton in his speech to the United Nations General Assembly put forward a comprehensive approach to the growing accumulation of fissile material from dismantled nuclear weapons and within civil nuclear programs. A factsheet issued by the White House described these plans in more detail, including the proposal for a "multilateral convention prohibiting the production of highly enriched uranium and plutonium for nuclear explosives purposes or outside of international safeguards."[76] Under this proposal, existing stockpiles of fissile material no longer required for nuclear weapon purposes would be submitted to inspection by the IAEA.

In October 1993, Russia followed suit in proposing negotiations on this issue at the Conference on Disarmament (CD).[77] In November 1993, the First Committee of the CD, for the first time in 15 years, by consensus passed a draft resolution on banning the production of fissile materials for weapons (see, e.g., Disarmament Newsletter, issued by the United Nations Center for Disarmament Affairs, Vol. 11, Number 2, November 1993, p. 9). Accordingly, the United Nations General Assembly (UNGA) adopted Resolution 48/75L calling for negotiations on this issue in December 1993. While Clinton's proposal identified HEU and plutonium as the fissile materials to be restricted, neither UN resolution specified the substances to be covered. Since tritium is not a fissile material, it is not an open question whether it will be included in production restrictions. Otherwise, the wording should be "nuclear weapons," "weapon-grade," or "fissile and fusionable" materials. However, tritium was mentioned a few times in the discussions (Disarmament Times, Vol. 16, No. 6, November 23, 1993, p. 1).

Significant progress in the cutoff issue will be required to comply with the "*principles and objectives*" laid down at the Review and Extension Conference of the NPT in 1995. It mentions a fissile material agreement as the next step after the CTBT. After the conclusion of negotiations for the CTBT at the CD in Geneva in 1996, it was expected that a cutoff agreement would be the next treaty to be negotiated there. In March 1995, the Canadian Ambassador Gerald E. Shannon reached consensus on a mandate for the CD to negotiate "a treaty banning the production of fissile material for nuclear weapons or other nuclear explosive devices" (CD/1299). Due to conflicting interests among various countries, the lowest common denominator reduced the scope of the Shannon mandate to the following three points:

1. the production of fissile material for nuclear weapons or other nuclear devices
2. the production of fissile material outside safeguards
3. fissile material released from military stocks into the civilian sector for inspection

Tritium is not covered by this mandate. Therefore, tritium will probably play no major role in the negotiations. It may come in by the backdoor. The advantage of this limited mandate is that the recognized nuclear weapons states have already established an almost complete de facto moratorium on production. Enshrining this into international law might not be too difficult to envision. The mandate explicitly states that it does not preclude any delegation from raising for consideration in the Ad Hoc Committee other issues such as stockpiles from past production of fissile materials and management of such materials.

Tritium control may be one of these topics that could be raised during the negotiations. A few delegations (India, Indonesia, Ireland, Mexico, South Africa, and others) showed a certain sympathy for some kind of tritium control within a cutoff. For example, South Africa in a working paper on "The possible scope and requirements of the Fissile Material Treaty (FMT)" (document number CD/1671, Geneva, May 2002) demands that "the production of tritium in civil nuclear reactors for use in nuclear explosive devices should be prohibited by the FMT."

A full-scale tritium production cutoff may be added to the fissile material ban. An Integrated Cutoff agreement is discussed in depth in Section 1.8.3. It should be noted that even if tritium does not play a major role, tritium production will inevitably be addressed. The important role of tritium production in a Fissile Material Treaty is due to the ease of converting a reactor from generating tritium to breeding plutonium. This causes either a need for verification at the tritium facility; otherwise a loophole or at least an uncertainty would remain in the plutonium production ban.

For more than three years, the CD failed to establish the Ad Hoc Committee. Eventually, in August 1998 the CD decided to establish a committee to negotiate a "nondiscriminatory, multilateral, and internationally, and effectively verifiable treaty banning the production of fissile material for nuclear weapons or other nuclear explosive devices" on the basis of the Shannon mandate of 1995. Since it is heavily disputed whether this treaty would address both nuclear disarmament and nonproliferation, it is remarkable that the CD established this Ad Hoc Committee under item 1 of its agenda entitled "Cessation of the nuclear arms race and nuclear disarmament." However, the negotiations did not start within the following five years.

U.S.–Russian bilateral agreements

The U.S. and Russia started detailed talks on fissile materials in May 1994. On June 23, 1994, Al Gore and Viktor Chernomyrdin signed an agreement on the shutdown of Russia's remaining three production reactors by the year 2000. In return, the U.S. agreed not to restart its own shutdown production reactors and to help the Russians find alternatives for generating the heat and electricity still provided by the production reactors. In addition, Russia agreed not to use newly produced plutonium for nuclear weapons. The U.S. and Russia also agreed to develop procedures necessary to assure compliance with the obligations of the agreement.

This agreement does not mention tritium. In its Annex, a list of plutonium production reactors is given which makes no mention of the shutdown K-Reactor at the Savannah River Plant and of the two still-operating light water reactors at Ozersk, named Lyudmila and Ruslan, each with a capacity of about 1000 MW_{th}, which are used to produce tritium and special isotopes, e.g., ^{238}Pu. This constitutes a severe loophole with respect to verifying compliance with this agreement, because these can serve as plutonium production reactors. In principle, all tritium production reactors can easily be used for plutonium production as well. The possibilities of exchanging the raw materials lithium and uranium depend on the configuration of the core and the design of the fuel and target elements. Besides exchanging the target materials, a slight reconfiguration of the core might be necessary. Normally, such a procedure is easy, because target elements are separated from fuel elements. Sometimes fuel and target material are integrated in the same elements, which would imply higher costs for converting a tritium breeding target to one that breeds plutonium. But there is no physical reason that impedes such a conversion. In fact, the K-Reactor, which is nearly identical to the plutonium production reactors L, P, R, and C at the Savannah River Plant, was used for the production of supergrade plutonium (3% in ^{240}Pu) for several years starting in 1983 (Cochran, 1987b).

The mission of this reactor was changed to tritium production after the shutdown of the C-Reactor in 1986, which was dedicated to tritium production. Coproduction of plutonium and tritium has been current practice. All U.S. production reactors have been shut down since 1988. Plans to restart the K-Reactor have been abandoned, but the new production line with commercial reactors is planned to start in full 2003 (see Section 2.7).

If under a "cutoff" agreement military tritium production reactors continued to operate and if the production reactors require HEU as driver fuel, this would cause additional complications for safeguarding military HEU stocks. However, it is very unlikely that a HEU production cutoff would be undermined because neither the U.S. nor Russia currently has HEU-consuming production reactors under operation, and because there is enough HEU in the military stocks from dismantled nuclear weapons as well as from reprocessing submarine fuel to run a reactor if needed.

1.8.2 The principle of reciprocity within a Fissile Material Treaty

The question addressed here is how disarmament can be integrated into the scope of a Fissile Material Treaty and what role tritium control could play for that purpose.

Overcoming the deep political differences with regard to the scope of the Fissile Material Treaty will necessarily require some compromise between states that emphasize disarmament and those that stress nonproliferation. For the success of negotiations on a cutoff agreement, this can mean following one of two approaches. Either a linkage is made between a cutoff mandate and separate negotiations on nuclear disarmament, or disarmament measures are integrated within the cutoff mandate.

The first approach might be realized by immediately starting negotiations towards a Nuclear Weapons Convention (NWC), which might serve as a framework for progress in both nuclear nonproliferation and disarmament in a reciprocal way.[78] Although many nonaligned states favor or even demand the first approach, only the second one is considered in this book. It belongs to the step-by-step approach which is complementary to the first one, the comprehensive approach.

The second approach aims at including special provisions in a cutoff agreement which has a disarming effect on nuclear weapon states. Frequently it is stated that the reduction of stockpiles from past production may serve this purpose. This is definitely the case if this reduction goes as far as eliminating any military nuclear material that is not placed in nuclear weapons of the active arsenal. Nevertheless, it is recommended here to consider tritium control as a provision which leads to qualitative disarmament with a similar but stronger effect compared to dealerting or sequestration of nuclear weapons.

Before discussing this concept of qualitative disarmament by means of a tritium production cutoff, another argument on including a disarmament provision in the Fissile Material Treaty is explained below to provide further motivation.

Disarmament provisions to balance the nonproliferation effect

In addition to satisfying the demand to link nonproliferation and disarmament measures, there is another strong argument for the integration of a disarmament measure into a cutoff agreement. It is generally accepted that a cutoff agreement should be nondiscriminatory. Some people take the narrow view that only obligations which are physically identical for all states would be nondiscriminatory. However, it has to be taken into account that identical provisions may have different impacts on the perceived national security interests of different groups of countries.

For example, a provision that all states should stop the production of fissile material for weapons purposes or outside of safeguards seems attractive due to the identity of demands on all states, but it has different impacts on various states. The nuclear weapons states do not need further production of fissile materials and in general already have a production moratorium in place. On the other hand, the threshold states may have the feeling that their nuclear options are significantly restricted and even more so if stocks from past production are included in the ban.

A mandate at the CD, which is in this sense discriminatory, and any agreement based on such a mandate will be hardly acceptable to those states that see themselves at a disadvantage. Therefore, it is imperative to search for reciprocal measures which have equivalent impact, especially on the five recognized nuclear weapons states and on the nuclear threshold states.

Thus, the only way of taking reciprocity seriously is to accept the different impacts and even to aim at a cutoff agreement that includes provisions that have a nonproliferation effect on threshold countries as well as a disarmament effect on nuclear weapon states. Any measure that places the unsafeguarded weapons-usable material and production facilities in nuclear threshold states under some sort of control serves by definition the goal of nonproliferation. Therefore, it is necessary to look for provisions which are clearly serving nuclear disarmament.

One could follow the argument that the reduction of stockpiles of nuclear-weapons-usable material is a disarmament measure. In fact, it would help to freeze the current active arsenal and it would make past progress in nuclear disarmament irreversible. It can even be counted as disarmament if one takes the view that warheads are completely disarmed only after their dismantlement and after transferring the material from the military realm to international safeguards.

Is there a chance to find reciprocal provisions by including past stockpiles in the cutoff agreement and how should these be defined? In particular, the question remains open whether unsafeguarded stockpiles in nuclear threshold states should be included in the ban as well.

Suitable equivalent steps might be easier to understand when the whole process from deep cuts down to a nuclear-weapons-free world is viewed. The nuclear threshold states had (or still have) a policy of nuclear ambiguity, which means that they neither deny nor confirm that they have nuclear weapons. This is continued by not publicizing how many they have. They are known to possess sufficient amounts of nuclear-weapons-usable material to produce a number of nuclear weapons. It is suggested here that in the process towards a nuclear-weapons-free world these states shall under no circumstances be officially recognized as nuclear weapons states. When these states join the nuclear disarmament process, they should reduce the upper limit of their stocks of nuclear-weapons-usable materials while the recognized nuclear weapons states further reduce the limits of their nuclear arsenals. In the last step towards eliminating nuclear weapons, the threshold states should surrender the remaining stocks of material while the nuclear weapons states surrender the remaining nuclear weapons. The complete surplus of nuclear material of the latter may be placed under control at an earlier stage.

From this logic it becomes apparent that the reductions of stocks of weapons-usable materials in the two different groups of countries are not equivalent. Materials in threshold states should be regarded as equivalent to nuclear weapon arsenals in recognized nuclear weapons states. Therefore, the current mandate at the CD for cutoff negotiations cannot be made nondiscriminatory simply by including into the agreement reductions of stocks from past production in all countries.

Qualitative disarmament effect with a weak timetable through a ban on tritium production

A different way to address disarmament within a cutoff treaty can be identified when considering qualitative disarmament of nuclear weapons (see Section 1.6.4). As a result of these considerations, a measure within an agreement on nuclear-weapons-usable materials that may have some impact on disarmament in the nuclear weapons states would be control of further production of tritium. This is because fresh supplies of this material may be necessary some time early in the 21st century in the case that nuclear disarmament stops to keep pace with the natural decay of this radioactive superheavy hydrogen isotope at 5.5% per year. Therefore, it is suggested here that a ban on tritium production in recognized nuclear weapons states should be taken as a measure that is equivalent to an appropriate control of fissile materials in threshold states.

In 1988, a similar suggestion was discussed in the U.S. with the intention of using the decay of tritium as a "forcing function" for nuclear disarmament (see Nuclear Control Institute [NCI] and American Academy of Arts and Sciences [AAAS], 1988, and Section 1.5, item 7). This proposal was rejected on good grounds mainly with the argument that it would be the tail wagging the dog. If there is an attempt to force nuclear disarmament by using the tritium decay as a forcing factor, the difficult negotiations about stockpile reductions are diverted. The proposal made here is significantly different from the one put forward earlier. If tritium is removed from a nuclear weapon, the weapon is qualitatively disarmed but it is still a nuclear weapon, and this process is reversible.[79] Tritium could be reinserted into the same weapon. Therefore, this process is similar to taking nuclear weapons off alert for a longer time. It would take longer to get them on alert again. It might even take one or several years to produce the required amount of tritium, depending on how long tritium was allowed to decay below the demand and depending on the capacity of dedicated production facilities.

Since it is considered here that only fresh production of tritium should be banned, the existing military inventories could be redistributed to the warheads in the active arsenal. It should be noted that the number of nuclear weapons affected by this approach would be very low at the beginning and would increase with time. No inspection of warheads would be required. It has only to be verified that no fresh tritium is produced. This is relatively straightforward, and the verification of a plutonium production ban would even be much easier if no reactor were reserved for tritium production (Kalinowski *et al.*, 1992 and Kalinowski, 1995b). This can even be done nonintrusively by remote monitoring.

When will the U.S. and Russia be prepared to agree to a step like the one proposed here? So far, the U.S. and Russia have not included nuclear warheads in any disarmament agreements. START I and II restrict only the number of delivery systems. Although it has been proposed that the dismantling of nuclear warheads should be included in START III, the Moscow Treaty of 2002 demonstrates that the U.S. and Russia prefer disarmament measures which avoid provisions that directly involve the warheads. Therefore, the proposal made here might be easier to accept than inspected sequestration (separate storage) of warheads and delivery systems.

The advantage of this approach is that a weak linkage is established between nonproliferation efforts directed against threshold states and disarmament measures addressing the recognized nuclear weapons states. This weak linkage avoids the seemingly unbridgeable gap between nuclear disarmament and nonproliferation. The linkage is weak because a ban on tritium production may never have a restricting effect on nuclear arsenals, provided that disarmament conducted in parallel keeps pace with the decay of tritium. If the Moscow Treaty is realized, no tritium production is necessary for at least the next two decades. Thus, the decay of tritium provides a soft and — if perceived to be necessary, a reversible — time-bound framework for nuclear disarmament and thus allows a compromise to be achieved between states of the nonaligned movement and the nuclear weapons states regarding such a demand.

1.8.3 The "integrated cutoff" (ICO)

As discussed in the previous subsection, a production ban on fissile materials would not really constitute significant progress in nuclear disarmament since a de facto production halt is already in place or planned for the near future, and plutonium and uranium practically do not decay like tritium does.

It is suggested here that the production cutoff should be supplemented by a tritium production cutoff. This combined cutoff has been named "integrated cutoff" (ICO) (Kalinowski *et al.*, 1992; Kalinowski and Colschen, 1995; and Colschen, 1995). In addition to the reasons given in subsection 1.8.2, a further reason for the ICO is that clandestine stockpiles of HEU and plutonium remain almost constant whereas tritium stockpiles decay at a significant rate. It is easy to imagine that a stockpile of up to perhaps 100 tonnes of uranium-235 could be hidden before a cutoff agreement became effective (von Hippel *et al.*, 1985). Consequently, it was pointed out earlier that it would be reasonable to include a limit on tritium stockpiles in a fissile materials production cutoff agreement: "This could be accomplished by an agreement between the two nations limiting their tritium production to below a common ceiling."[80]

Another reason to include tritium in a cutoff is that an integrated cutoff would be easier to verify than a pure fissile materials cutoff, mainly because plutonium could otherwise be produced clandestinely in a nuclear facility which is declared to produce tritium.[81]

In 1988 it became public that the U.S. encountered difficulties in producing tritium for military purposes. The last military production reactor at Hanford (the N-Reactor) had been shut down in January 1987. In April 1988, severe problems with safety and radioactive contamination led to the shutdown of the remaining operable production reactors at the Savannah River Plant near Aiken, South Carolina, including the dedicated tritium production reactor (the K-Reactor).[82]

The necessity of deciding about huge investments for new tritium-producing facilities in the U.S. triggered a discussion about a cutoff of tritium. One idea was to pursue an indirect strategy for nuclear disarmament by making use of the so-called tritium factor (see Section 1.6.4 and Section 1.8.2, item 'qualitative disarmament effect' as well as Wilkie, 1984; Mark, 1988; Sutcliff, 1988; and Nuclear Control Institute [NCI] and American Academy of Arts and Sciences [AAAS], 1988). The dominating effect of tritium decay on disarmament can be shown by mathematical modelling (Scheffran, 1989).

Therefore, including tritium in the cutoff agreement would constitute a stronger commitment by the nuclear weapons states towards complete nuclear disarmament. Sooner or later, depending on the pace of the nuclear disarmament process, there will come a date at which all reserves of tritium have vanished due to its decay. If implementation of the Moscow Treaty proceeds as planned, there will be no need to resume tritium production in the U.S. or in Russia for the foreseeable future.

Although such proposals sound rather attractive because of their apparent simplicity, it is unlikely that a technically induced mechanism instead of traditional nuclear disarmament negotiations or unilateral measures will be politically acceptable. Consequently, this proposal has never reached the agenda of policy makers in the U.S. or elsewhere. A more realistic approach is the integrated cutoff, which can be negotiated in a way that takes into account that fresh tritium supplies may be necessary after some decades.

The U.S. will not need any new tritium before 2020 and probably far beyond that date.[83] Russia probably still has tritium production reactors in operation. The Russian demand for tritium is not as well known as that of the U.S., but it can be assumed that the situation is similar. The U.S. policy is to have the option to resume tritium production for nuclear weapons whenever it is considered necessary. In fact the start of a new dedicated tritium production has been decided. The ICO would guarantee that time is bought by postponing new production activities.

The *goal* of the ICO is the nonavailability of **fresh tritium supplies** for nuclear weapons programs as a means of avoiding the vertical proliferation of states that possess nuclear weapons or weapons capability and to pave the way to complete nuclear disarmament, i.e., the denuclearization of those states. Only the recognized and de facto nuclear weapons states are potential member states of the ICO.

The four tritium-related key *rules* of the ICO are the following:[84]

1. No tritium will be produced for nuclear weapons purposes.
2. All military facilities for the production of tritium are shut down and kept on a status which is comparable for all states.

3. No new facilities for the production of tritium will be constructed or developed, including new tritium production technologies such as the accelerator technology.
4. No civilian facilities will be converted to military facilities or made use of for military purposes, and no tritium produced in civilian facilities will be transferred to military uses.

The main *advantages* of an ICO in comparison to a "fissile material cutoff" are:

- Compared to the "cutoff" proposal, the ICO would constitute a **stronger commitment** by the nuclear weapons states to complete nuclear disarmament and it is more suited to satisfying the demands of nonnuclear weapons states.
- The ICO is less discriminatory than a Fissile Material Treaty because a measure with a possible qualitative disarmament effect is included. If the Moscow Treaty is not progressively implemented and then followed by further reductions, within a few years the stockpile would start to be qualitatively disarmed by a yield reduction, which is due to the decay of the tritium inventory below the arsenal's demand (see Section 1.6).
- If the implementation of the Moscow Treaty proceeds as planned, there will be no need to resume tritium production in the two nuclear weapon states party to the treaty for more than 20 years (see above). Therefore, any possible asymmetries regarding the impact of tritium shortages on the nuclear arsenal of the two countries will be irrelevant for this period of time.
- **Verification** of this "zero approach" would be easier and less intrusive.
- Substantial **cost-saving effects** on two levels would result from an ICO. First, there are the costs to maintain and possibly even to build new tritium production reactors.[85] Second, if verification were confined to military facilities, there would be additional costs involved to verify a "cutoff," which would be unnecessary within the ICO.

Despite these advantages, in the past tritium has rarely been addressed in discussions about a cutoff and was almost never mentioned in official documents.[86] One reason for that is that the significance of tritium for vertical prolifertion and nuclear disarmament has been underestimated. Another reason is that it is generally believed that the verification of a tritium cutoff is not feasible. This monograph proves the contrary.

Two tasks for *verification of nonproduction* of tritium in military facilities can be distinguished:

1. Verification of inactivity of production facilities (see Section 3.3.2):
   As long as former military production facilities are not dismantled, it has to be verified that they are not operating. The status (dismantled, mothballed/standby, operating) of a known production facility could be verified by remote sensing using national technical means or, for example, by a future

international satellite verification agency. In addition, special on-site inspection activities might be agreed to verify that measures are taken which disable a quick restart. The status of "cold standby" has to be defined in technical details.

2. Detection of clandestine facilities and activities (see Section 3.6):
   It has to be verified that no clandestine facilities are constructed and operated. To some degree, such activities could also be detected by remote sensing activities (see, e.g., von Hippel and Levi, 1986 and Jasani and Stein, 2002). Detection could trigger further on-site inspections to scrutinize the alleged illegal activities.

In case of more stringent requirements on verification, the *inspection of civilian facilities* would be included in the verification approach. This would be added at a later stage and would constitute the link of an international agreement on horizontal nonproliferation of tritium (see Section 1.7.2) with the integrated cutoff. Again, two different tasks can be distinguished:

- Detection of breeding activities (see Section 3.3.3):
  It has to be verified that no nuclear energy is diverted from civilian facilities to produce tritium for nuclear weapons purposes.

- Verification of nonremoval (see Section 3.4):
  It has to be verified that no tritium from civilian sources is diverted to nuclear weapons purposes. Part of this task is to determine the baseline of tritium produced inadvertently (see Section 3.3.4). Tritium accountancy would be required for all civilian stocks and transfers of quantities above a certain limit.

These two verification tasks would be based on control procedures at civilian nuclear facilities at least in nuclear weapons states, but possibly in all countries which are party to the treaty. Whether civilian facilities would be included either in a cutoff agreement for fissile materials or in an integrated cutoff including tritium depends on the political will to pay for the verification procedures. If complete coverage of all civilian production and diversion paths for these materials was required, IAEA-type safeguards would have to be implemented and supplemented by additional measures to verify tritium nonproduction and nonremoval. Nevertheless, the contention of this monograph is that a comprehensive verification goal regarding tritium can be achieved at reasonable cost (see Section 3.8).

1.9 Endnotes

1. For general information about tritium, its chemistry and various applications see Evans (1974) and Fiege (1992).

2. A comprehensive overview of tritium sources and radiological consequences is given in Martin (1982) and Moghissi and Carter (1973). Broad information about the management of tritium at nuclear facilities can be found in International Atomic Energy Agency IAEA (1984) and IAEA (1991).

3. For more details on measurement technology see, e.g., United States National Committee on Radiation Protection (1976), and on health and environmental aspects see, e.g., Straume (1993).

4. The dramatic increase in tritium demand has been noticed by Morland (1980).

5. According to CFFTP (1988), the yearly consumption of tritium worldwide stabilized at about 200 g. Other sources estimate the civilian use at a total of 500 to 1000 g per year, e.g., Cochran (1987), but this is definitely too high. It seems that nobody in the world has a complete overview of tritium sales. Probably the best available survey on international trade of tritium is given in Colschen *et al.* (1991). Imported quantities of 14 countries — including by far the then-largest importers Canada, Switzerland, and the U.K. — for various periods in the 1980s are listed in that documentation. The average yearly import of these countries amounted to 78,000 TBq or about 217 g. Countries not included had yearly imports and exports which are less or much less than one gram. Some tritium appears more than once in the statistics because it is reexported, e.g., after manufacturing to radio-luminous paint.

6. During 1982 and 1988, an average of 58,000 TBq per year was supplied by the U.S. (United States Nuclear Regulatory Commission, 1989).

7. See United States Nuclear Regulatory Commission (1989). One source states a price of about DM 10,000 per gram (Lieser, 1980). Another source gives a price of $7,500 per gram at a purity of 94% (Seifritz, 1984).

8. There are indications that the U.K. is unable to satisfy the commercial tritium market from its military production and might even have difficulties in meeting its military needs without imports from the U.S. From 1983 to 1988, British companies received extraordinarily large amounts of tritium from the U.S., officially for commercial applications (120 g in 1983, 50 g in 1984, 75 g in 1985, 71 g in 1986, 100 g in 1987, and 300 g in 1988 according to the United States Nuclear Regulatory Commission, 1989).

9. Alkor Technologies, a private company in St. Petersburg, considers bulk exports of tritium. The military facility at Ozersk (former Chelyabinsk-65) together with the state-run Radium Institute in St. Petersburg are planning to manufacture luminescent signs containing tritium.

10. Some tritium may be available from detritiation of aqueous waste streams from the shut down reprocessing plant at Mol.

11. Tritium recovery from the heavy water of the high flux reactor at the Laue Langevin Institute in Grenoble was able to separate up to 16 g per year. This facility was operating from 1972 and is currently on stand-by.

12. For the facilities with the largest inventories, see Table A.12 in Appendix A.

13. For a technical description of these neutron generators, see the next section.

14. These estimated quantities include not only the amount of tritium actually inserted in the warhead but also its share of the working inventory in the production and replenishment "pipeline."

15. One neutron generation is called a shake and lasts for approximately 10 ns.

16. The more neutrons offered in the first "shake," the higher the probability that a chain reaction is started.

17. The target consists of a copper plate, which is coated with zirconium or titanium. Tritium is absorbed ("gettered") by this thin layer like water by a sponge. The production technology for such targets is described in Wilson (1954).

18. The most frequently cited amount is 4 g. It is an overestimation and may have been derived by simply dividing the total tritium inventory by the number of warheads in stockpile. The best available estimate of tritium production made by Thomas Cochrane is based on publications of routine atmospheric releases of tritium at the production site. Accordingly, the total stock of tritium in the military arsenal of the U.S. was (70±25) kg in 1984 (Cochran, 1987). In the same year, the U.S. had 23,500 nuclear weapons in the arsenal. Accordingly, the average amount of tritium per warhead was 3.0 g. This is the upper bound for the average inventory of tritium that is either really placed in the weapons or remains in the production and refreshing pipeline, because the U.S. DOE always tries to keep an emergency surplus.

19. The physical principle of boosting is explained in Gsponer and Hurni (1998).

20. If due to a problem like preinitiation of the chain reaction only a limited unboosted yield of 0.1 kt is achieved, the boosted yield can be two orders of magnitude higher (Gsponer and Hurni, 1998).

21. In 1953 the Soviet Union tested Joe 4, a single-stage boosted fission weapon with a yield in the 400 kt range (Cochran and Norris, 1993, p. 20).

22. Allowing for some tritium in the production pipeline and some reserved for neutron bombs, it can be assumed that some 50 to 70 kg were available in 1988 for about 23,000 nuclear warheads. Thus, between 2 and 3 grams are used on average for boosting. The actual content is believed to vary significantly between different types of warheads.

23. This was concluded from a nuclear weapon test conducted in 1962 by Lawrence Livermore Laboratory. The yield of a one-year-old W-45 warhead reached just half of the expected value (Hansen, 1988, p. 184). This radical decrease was not predictable and is not easy to understand. Tritium's decay product helium-3 is a strong neutron absorber and acts as a neutron poison. Assuming that pure tritium had been inserted initially, the helium-3 content after about one year could not have been much higher than 5.5%. It is not obvious how such a small quantity of helium-3 could suffice to explain the reduction of the total expected yield by 50%. Another possibility is that the tritium has dropped by decay below a critical quantity.

24. In Margeride (1978) it is estimated that at least 7.2 g of tritium is needed for one neutron bomb.

25. For details, see Sections 1.3.1 and 1.3.2.

26. This chronology is a slightly modified translation from Kalinowski and Colschen (1993).

27. It was jointly organized by IANUS (Interdisciplinary Research Group in Science, Technology and Security), INESAP (International Network of Engineers and Scientists Against Proliferation), and UNIDIR (United Nations Institute for Disarmament Research).

28. Since the military production rate is not known, the inventory change due to radioactive decay is used as the comparative quantity. See Section 2.6.9.

29. In 1970, it was suggested at a Pugwash conference that tritium should be included in IAEA safeguards so as to obtain adequate assurance that non-fission-triggered thermonuclear weapons cannot be produced by nonnuclear-weapons states which are parties to the NPT. This proposal was based on the assumption that it would be possible to build a laser-triggered thermonuclear weapon (Olgaard, 1971). This, however, is highly unlikely.

30. A German chemist and tritium expert who organized in the mid-1980s the illegal transfer of tritium to Pakistan from the German company NTG stated in the trial that he was not aware of the fact that gram amounts of tritium would be of relevance for nuclear weapons. At a workshop on tritium as fusion fuel, the participants were surprised and startled by a presentation on nuclear weapons' use of tritium and possible control measures. See Kalinowski (1990).

31. For an extensive analysis of the civil/military ambivalence of tritium, see Kalinowski and Colschen (1993).

32. For example, the German Bundeswehr renounced the procurement of radio-luminous paint for radiation protection purposes with one exception (Heyden, 1991). According to Schroeder, Bundesministerium für Verteidigung (1992), this exception is capsuled sources, e.g., tritium-containing radio-luminous paint and ionization detectors. Tritium will be substituted and old stocks will partially be eliminated.

33. Note that the accountable quantity within an MBA is one-tenth of the accountable quantity for the whole facility when reporting to the DOE (see previous section) and that accountable quantities are larger than those quantities for which a license is required.

34. Table IV 1 of the appendix states 5×10^6 Bq for tritium.

35. For more details see Colschen *et al.* (1991), pp. 40–42.

36. All tritium, regardless of form or chemical composition, that is no longer usable in experiments is designated as *scrap*. The scrap material must be evaluated by the DOE before it may be qualified as waste (Wall and Cruz, 1985).

37. This case happened in July and August 1988 when two British companies received in total 5 g of tritium less than they had ordered (Broad, 1989).

38. However, it was totally unrealistic to assume that this much money could be earned by selling the tritium on the civilian market, since these 2.5 kg per year are about six times as much tritium as the whole world market presently demands each year. See Figure 1.2.

39. For further details see Colschen *et al.* (1991), pp. 32–33.

40. Export regulations and special provisions regarding tritium are described for a number of countries in Nuclear Energy Agency (OECD) (1988).

41. The limits are quoted in giga-Bequerel (GBq) with 1 GBq = 10^9 Bq. For comparison, roughly 3.6×10^{14} Bq = 360,000 GBq are one gram. Some of the values stated in Table 1.4 may have changed in the meantime, because the regulating authorities of 6 out of 20 countries stated that their export regulations were under revision at the time of the inquiry, and four countries answered that they believed the tritium regulations were inadequate (Colschen *et al.* (1991)).

42. The remaining six CoCom member states were not included in the review. See Colschen *et al.* (1991).

43. In the joint working committee, Canada is represented by the Atomic Energy Control Board (AECB) and Europe among others by EURATOM.

44. For a detailed analysis of accountancy procedures at these facilities, see Section 3.4.

45. See proposed international tritium control agreements in Sections 1.6, 1.7 and 1.8. For more elaborated scenarios see Colschen and Kalinowski (1994) and Colschen (1998). The proposal to put a limit on tritium production at declared production facilities was investigated by Stern (1988).

46. In fact, this is a bilateral agreement in which the receivers are member states of the supervising agency. See Section 1.4.3.

47. An International Tritium Control System (ITCS) has been proposed. See Section 1.7.2 as well as Kalinowski and Colschen (1993); Colschen and Kalinowski (1994); and Colschen (1998).

48. Dealerting of nuclear weapons is proposed in various studies such as the Report of the Canberra Commission on the Elimination of Nuclear Weapons (August 1996). The concept is explained by Blair (1995) and von Hippel (1997).

49. Part of this section was published in Kalinowski (1995).

50. Data are taken from Cochran *et al.* (1984); Hansen (1988); Norris (1993); and Norris and Arkin (1994).

51. In this monograph, the yield-to-weight ratio is defined as the total explosive yield divided by the total weight of the complete weapon. The mass of the fissile and thermonuclear fuel core is not of interest in this context and is usually not known to the public.

52. For an explanation of boosting see Section 1.3.2.

53. This test was carried out in 1952 (Hansen, 1988). Its core, which was designed by Theodore B. Taylor, was made of "oralloy." In the 1960s, Ted Taylor became an opponent of nuclear weapons and is now engaged in activities to promote a nuclear-weapons-free world.

54. See Hansen (1988). If due to a problem like preinitiation of the chain reaction only a limited unboosted yield of 0.1 kt is achieved, the boosted yield can be two orders of magnitude higher (Gsponer and Hurni, 1998).

55. Part of this section is an updated version of an assessment in Kalinowski (1995).

56. Part of this section has been published in Kalinowski (1995).

57. The observation that the aging of tritium qualitatively disarms nuclear arsenals has been noted much earlier (Wilkie, 1984) and has been mathematically modeled in Scheffran (1989).

58. In a special way this proposal has been mentioned earlier by Albright and Zamora (1989). They referred to governmental officials who have suggested several steps to extend the supply of tritium, among others by eliminating the yield options in some nuclear weapons. The yield can be selected by inserting a number of tritium reservoirs in the warhead (see Section 1.3.2).

59. The Integrated Cutoff (ICO) proposed in Section 1.8.3 bears an asymmetry in that one country will run out of tritium earlier than the other. This is avoided by complete elimination of tritium.

60. There is some possibility that the weight of heavier types of warheads could be reduced by removing certain components, especially the second stage, leaving back the boosted primary with a maximum yield of about 10 kt. Thus this approach might have a similar effect as tritium elimination, but it would be more intrusive to find out the relevant weight limit which would probably be different for different warhead designs.

61. For a boosted nuclear weapon the density of tritium and deuterium is desired to be maximum in the center of the explosion after a considerable amount of energy has been released from fission. See Hansen (1988), p. 29. To achieve this, the correct timing of the injection of the tritium and deuterium gas may be critical. The movement of gas particles takes a long time in comparison to the typical time frames of a nuclear chain reaction measured in shakes (one shake equals 10 ns) and the gas has to be injected in due time before the implosion.

62. It took the U.S. 70 months for the first step and another 18 for the second; the corresponding numbers for the other NWS are: Soviet Union 48 and 27, U.K. 55 and 6, France 79 and 23, and China 19 and 13 months.

63. According to Rodney Jones (1983), a small nuclear force is something more than a nuclear explosive capability. He argues that a minimum number of warheads, perhaps as few as half a dozen, would be regarded as the minimum for a country in order to have some redundancy as insurance against unreliability or as a poststrike reserve to deter retaliation.

64. In 1978, the atomic research center at Bhabha reported on measurements of lithium enrichment with mass spectroscopy. See Stern (1988), p. 105.

65. See United Nations Special Committee (1991). The natural abundance of lithium-6 is 7.5%.

66. Quoted in the Arms Control Reporter 10,6 (1991) 602.B.197.

67. See Kalinowski and Colschen (1995); Colschen and Kalinowski (1994); and for regime theoretical background information, Colschen (1998).

68. This also includes tritium transfers to or between recognized nuclear weapons states. It should be stressed that this rule goes beyond the rules of the NPT regarding the transfer of nuclear materials.

69. In combination with the following rule this means that tritium exports require "full-scope safeguards."

70. Compare definition given in Art.98 (2) O of the Verification Agreement dated 5 April 1973 and in Art. XX IAEA Statute.

71. See Hoodbhoy and Kalinowski (1996). This subsection draws from that publication.

72. Three reprocessing facilities with a total capacity of 275 t heavy metal per year are in operation in India.

73. A very comprehensive overview of the cutoff topic is given in Schaper (1997).

74. Deep reductions in strategic arsenals down to 3000–3500 by end of 2003 were agreed in 1993.

75. For example, a bill for a "Global Nuclear Weapons Material Control Act" was introduced into the U.S. Senate on July 26, 1991, by Senator Timothy E. Wirth and others as an amendment to the FY 1992 Defense Authorization Bill [S.1576]. It urges the President to negotiate cooperative inspection measures to ensure that the Soviet Union promptly reciprocates the U.S. halt in the production of plutonium and highly enriched uranium for weapons. It also urges the President to further the U.S. nonproliferation goals by working with the Soviet Union to extend a ban on nuclear weapons material production to all nations. Also, Representatives Les Aspin and Mike Synar introduced H.R.3764, the "Nuclear Weapons Material Production Termination Act," into

the House of Representatives on November 8, 1991. This bill would have cut off all funds for the production of plutonium, highly enriched uranium, and tritium for weapons after August 1, 1992.

76. This is reproduced, e.g., in Trust and Verify No. 41, October 1993.

77. The CD is the only multilateral body for negotiations about disarmament and arms control in which states of all regions of the world are represented.

78. A draft model NWC was drafted by NGOs and launched on April 7 in New York. The drafting group was convened by the Lawyers Committee on Nuclear Policy (LCNP), and input on physical and technical issues was provided by the International Network of Engineers and Scientists Against Proliferation (INESAP). The model NWC was submitted to the UN Secretary General and became an official UN document with the number A/C.1.52/7.

79. A warhead without tritium is still a nuclear weapon with a significant yield because the unboosted fission primary still works with an explosive yield of around one kiloton TNT. In case of warheads with selectable yield, the lowest yield which is of military interest is achievable without tritium. See Section 1.6.2.

80. Quotation from von Hippel and Levi (1986). It should be noted that a limit on stockpiles is not identical to a production cutoff, because it may allow further production to compensate for tritium lost by decay.

81. For the impact of tritium production on the verification of a fissile material cutoff, see Section 3.3.

82. See, e.g., Albright and Beard (1989). Initially, the DOE planned to restart the K-Reactor in 1992 at reduced power for 12 months to demonstrate that tritium production could be resumed. On June 8, 1992, the K-Reactor achieved criticality for the first time since 1988, but was soon shut down again. In the meantime it was decided that this reactor would be shut down for good. See Section 2.7.

83. This is the most conservative calculation of available tritium supplies. See Section 2.7 and Nuclear Control Institute (NCI) (1989), which argues that the post-START II arsenals could be maintained for 35 to 40 years without new tritium production.

84. The ICO would be composed of a complex structure of rules regarding the cutoff of the production of plutonium, HEU, and tritium. The rules specified here apply to the cutoff of tritium alone. Some of these rules are also valid for plutonium and HEU. Others, not mentioned in this book, apply solely to HEU and plutonium, but not to tritium. For example, plutonium and HEU from dismantled warheads would be placed in internationally controlled stockpiles, whereas tritium will be recycled. Since no fresh tritium is added to the weapons programs, the tritium inventories of the states party to the treaty would shrink by 5.5% annually. A different approach that would eliminate

all tritium from the nuclear arsenals to achieve the immediate effect of a significant yield reduction is discussed in Section 1.6.3.

85. In case of a limited agreement and the resumption of military tritium production after the agreement expires and is not extended, the ICO would at least have bought time for further investment decisions.

86. One exception is a working paper presented on April 20, 1982, by Mexico's Ambassador Garcia-Robles, chairman of the UN Committee on Disarmament Ad Hoc Working Group on a Comprehensive Program of Disarmament. He presented a proposal for a partial text of a treaty on General and Complete Disarmament. A cutoff agreement is suggested for the initial phases of the proposal, which included for the first time a cutoff for the production of "fusionable" material, placing it under IAEA safeguards.

References

Arms Control Reporter (1994) The End of CoCom. April 1, 250, 30, Cambridge.

Albright, D. (1993) Slow but steady. *The Bulletin of the Atomic Scientists*, July/August, 5–6.

Albright, D. (1994) South Africa's Secret Nuclear Weapons. Institute for Science and International Security Report, May, Washington.

Albright, D. and Beard, J. (1989) The Tritium Follies. *The Bulletin of the Atomic Scientists*, November, 42–45.

Albright, D. and Paine, C. (1988) A Case against Producing Nuclear Material. *The Bulletin of the Atomic Scientists*, January/February, 46.

Albright, D. and Taylor, T.B. (1988) A Little Tritium Goes a Long Way. *The Bulletin of the Atomic Scientists*, January/February, 39.

Albright, D. and Zamora, T. (1989) India, Pakistan's Nuclear Weapons: All Pieces in Place. *The Bulletin of the Atomic Scientists*, 45, No. 5, 20–26.

Anonymous (1959) Mehr Tritium verfügbar [More Tritium available]. *Die Atomwirtschaft*, June, 268.

Barker, B., *et al.* (1998) Monitoring Nuclear Tests. *Science*, 281, 1967/8.

Bundesgerichtshof (1992) Entscheidung vom 31 Januar 1992 [Decision of January 31, 1992]. Aktenzeichen 2 StR 250/91, Karlsruhe.

Blair, B. (1995) *Global Zero Alert for Nuclear Forces*. Brookings Institution.

Blair, B., Feiveson, H.A. and von Hippel, F. (1997) Taking Nuclear Weapons off Hair-Trigger Alert. *Scientific American*, November, 74–81.

Bonizzoni, G. and Sindoni, E. (eds) (1990) Tritium and Advanced Fuels in Fusion Reactors. Proc. Course and Workshop. International School of Plasma Physics 'Piero Caldirola,' Varenna, Sept. 6–15, 1989, Bologna.

Broad, W.J. (1989) U.S. halts sale of tritium after loss of enough to make a nuclear bomb. *The New York Times*, July 26.

Canadian Environmental Law Association (1986) Preliminary Submission by the CELA to Ontario Hydro regarding the potential use, sale and export of tritium by Ontario Hydro. Toronto.

CFFTP (1988) Tritium Supply for Near-Term Fusion Devices. CFFTP-G-88024, May.

Cochran, T.B., Arkin, W.M. and Hoenig, M.M. (1984) *Nuclear Weapons Data Book. Volume 1: U.S. Nuclear Forces and Capabilities.* Cambridge.

Cochran, T.B., *et al.* (1987a) *Nuclear Weapons Data Book. Vol. 2: U.S. Nuclear Warhead Production.* Cambridge.

Cochran, T.B., *et al.* (1987b) *Nuclear Weapons Data Book. Vol. 3: U.S. Nuclear Warhead Facility Profiles.* Cambridge.

Cochran, T.B. and Norris, R.S. (1993) Nuclear Weapons Databook: Russian/Soviet Nuclear Warhead Production. Working Paper NWD 93-1. Washington.

Colschen, L.C. (1998) Die Internationalisierung der Tritiumkontrolle als Baustein des Nichtverbreitungsregimes für Kernwaffen. Bedingungen, Einflußfaktoren und Folgen. Ph.D. thesis accepted by the Free University Berlin 1997. Shaker Verlag, Aachen.

Colschen, L.C. and Kalinowski, M.B. (1994) Can International Safeguards be Expanded to Cover Tritium? Paper IAEA-SM-333/27. In *Proc. IAEA Symposium on "International Nuclear Safeguards 1994: Vision for the Future"*, Vienna, 14–18 March, Proceedings Series No. 945, Vol. 1, 493–503.

Colschen, L.C., Kalinowski, M.B. and Vydra, J. (1991) Comparative Documentation. National Regulations of Accounting for and Control of Tritium. IANUS-2/1991, Darmstadt.

Craig, H. and Lal, D. (1961) The production rate of natural tritium. *Tellus*, 13, 85–105.

Crawford, M. (1989) Accelerator Eyed for Warhead Tritium. *Science*, January, 469.

Desroches, J. (1973) Peintures luminescentes et lampes auto-luminescentes. *Bull. d'Informations Scientific et Techniques*, No. 178, Fevrier, 69–72.

Deutscher Bundestag (1990) Bericht des 2. Untersuchungsausschusses der 11 Wahlperiode, Drucksache 11/7800, Bonn.

Donnelly, W.H. (1989) Nuclear Arms Control. Disposal of Nuclear Warheads. Congressional Research Service. Issue Brief IB88024 updated May 24.

Ehrlich, P.R., *et al.* (1983) Long-term biological consequences of nuclear war. *Science*, 222, 1293–1300.

EURATOM and Canada (1991, May) Agreement in the form of an exchange of letters between the European Atomic Energy Community (Euratom) and the Government of Canada amending the Agreement between the European Atomic Energy Community and the Government of Canada for cooperation in the peaceful uses of atomic energy of 6 October 1959. EL/CEEA/CDN/GER. Reproduced in *Official Journal of the European Communities*, No. C 215/5, August 17.

Evans, E.A. (1974) *Tritium and its Compounds.*

Feld, B.T., *et al.* (eds) (1971) Impact of New Technologies on the Arms Race. In *Proc. of the 10th Pugwash Symposium*, Wingspread, Racine, Wisconsin, June 26–29, Cambridge.

Ferguson, C.D. (1999) TVA gets the nod. *The Bulletin of the Atomic Science*, March/April, 12–14.

Fiege, A. (1992) Tritium. Projekt Kernfusion, Juli, Karlsruhe.

Gsponer, A. (1984) La bombe a neutrons. *La Recherche*, No. 158, September, 1128–1138.

Gsponer, A. and Hurni, J.-P. (1998) Fourth Generation Nuclear Weapons. The Physical Principles of Thermonuclear Explosives, Intertial Confinement Fusion, and the Quest for Fourth Generation Nuclear Weapons. INESAP Technical Report No. 1, Fourth printing, May.

Hansen, C. (1988) *U.S. Nuclear Weapons. A Secret History*. Arlington, Texas.

Heusinger, H. and Rau, H. (1961) Lumineszenzanregung mit Tritium-Beta-Strahlung. *Kerntechnik*, 3, No. 3, 67–70.

Heyden, J. (1991) Wehrtechnik in einer sich wandelnden Umwelt. *Jahrbuch der Wehrtechnik*, 20, 12–20, Bonn.

Hibbs, M. (1989a) German firms exported tritium purification plant to Pakistan. *Nuclear Fuel*, February, 6–7.

Hibbs, M. (1989b) Probe of German firms' exports to India, Pakistan expected to move to U.S.. *Nuclear Fuel*, August, 4–5.

Hibbs, M. (1991) Nonproliferation bill introduced to restrict export of dual-use items. *Nuclear Fuel*, 16, July.

von Hippel, F. (1997) De-alerting. *The Bulletin of the Atomic Scientists*, 35, May/June.

von Hippel, F., Albright, D. and Levi, B. (1985) Stopping the Production of Fissile Materials for Weapons. *Scientific American*, 253, No. 3, 26.

von Hippel, F., Feiveson, H.A. and Paine, C.E. (1987) A Low-Threshold Nuclear Test Ban. *International Security*, 12, No. 2, 135–151.

von Hippel, F. and Levi, B. (1986) Controlling Nuclear Weapons at the Source: Verification of a Cut-Off in Production of Plutonium and Highly Enriched Uranium for Nuclear Weapons. In *Arms Control Verification*, Tsipis, K., *et al.* (eds).

Hoodbhoy, P. and Kalinowski, M.B. (1996) The Tritium Solution. *The Bulletin of the Atomic Scientists*, 52, July/August, 41–44.

Huaqiu, L. (1988) *China and the Neutron Bomb*. Occasional Paper of the Center for International Security and Arms Control, June, Stanford University.

International Atomic Energy Agency (1962) Tritium in the physical and biological sciences. In *Proceedings of the Symposium on the detection and use of tritium in the physical and biological sciences*, May 3–10, 1961, Vienna.

International Atomic Energy Agency (1967) Radiation Protection Standards for Radioluminous Timepieces. Recommendations of the European Nuclear Energy Agency and the International Atomic Energy Agency. Safety Series No. 23, Wien.

International Atomic Energy Agency (1979) Behaviour of Tritium in the Environment. In *Proc. Symp. San Francisco*, October 16–20, 1978, IAEA-SM-232/49, Vienna.

International Atomic Energy Agency (1984) Management of Tritium at Nuclear Facilities. Final Report. Technical Reports Series No. 234, Vienna.

International Atomic Energy Agency (1991) Safe Handling of Tritium. Review of Data and Experience. Technical Report Series No. 324, Vienna.

Jasani, B. and Stein, G. (2002) (eds) *Commercial Satellite Imagery. A Tactic in Nuclear Weapon Deterrence*. Springer-Verlag, Berlin etc.

Jones, R.W. (1983) *Small Nuclear Forces*. Washington.

Jones, S. and von Hippel, F. (1998) The Question of Pure Fusion Explosions Under the CTBT. *Science and Global Security*, 7, No.1, 1–22.

Jones, S., Kidder, R. and von Hippel, F. (1998) The Question of Pure Fusion Explosions Under the CTBT. *Physics Today*, September, 57–59.

Kalinowski, M.B. (1990) Nuclear Weapons Uses of Tritium and Multilateral Control Measures. In *Tritium and Advanced Fuels in Fusion Reactors*. G. Bonizzoni, E. Sindoni (eds), Proc. Course and Workshop, International School of Plasma Physics 'Piero Caldirola,' Varenna, Sept. 6–15, 1989, Bologna.

Kalinowski, M.B. (1993) Uncertainty and Range of Alternative in Estimating Tritium Emissions from Proposed Fusion Power Reactors and their Radiological Impact. *Journal of Fusion Energy*, 12, 391–395.

Kalinowski, M.B. (1995a) *The impact of complete elimination of tritium on a nuclear arsenal*. Appendix A to Kalinowski and Colschen (1995), 187–196.

Kalinowski, M.B. (1995b) The Role of Tritium Within a Verified Cutoff of Fissile and Fusionable Materials Production. In *Against Proliferation — Towards General Disarmament* (pp. 61–64), W. Liebert, J. Scheffran (eds), Proceedings of the First INESAP Conference at Mülheim, August 1993, Agenda Verlag, Münster.

Kalinowski, M.B. and Colschen, L.C. (1993) Lassen sich ziviler und militärischer Kontext kernwaffenrelevanter Materialien trennen? IANUS Working Paper 7/1993, Darmstadt.

Kalinowski, M.B. and Colschen, L.C. (1995) International Control of Tritium to Prevent its Horizontal Proliferation and to Foster Nuclear Disarmament. *Science and Global Security*, 5, No. 2, 131–203.

Kalinowski, M.B., Colschen, L.C. and Leventhal, P. (1992) Why and How Tritium Should be Considered Under a Verified Cutoff of Fissile Materials Production. In *42nd Pugwash Conference on Science and World Affairs*, September 11–17, Berlin.

Kobisk, E.H., *et al.* (1989) Tritium-Processing Operations at the Oak Ridge National Laboratory with Emphasis on Safe-Handling Practices. *Nucl. Instr. Methods*, A282, 329–340.

Krejci, Z. (1979) Tritium pollution in the Swiss luminous compound industry. In *Behaviour of Tritium in the Environment*, Proc. Symp. San Francisco, October, 16–20 1978, pp. 65–78, IAEA-SM-232/49, Vienna.

Lieser H.K. (1980) *Kernchemie*. 2nd edition.

Makhijani, A. and Zerriffi, H. (1998) *Dangerous Thermonuclear Quest: The Potential of Explosive Fusion Research for the Development of Pure Fusion Weapons*, Institute for Energy and Environmental Research, Takoma Park.

Margeride, J.B. (1978) Qu'est ce que la bombe *à* neutron? *Defense nationale*, 34, December, 95–112.

Mark, J.C., *et al.* (1988) The Tritium Factor as a Forcing Function in Nuclear Arms Reduction Talks. *Science*, 241, 1166.

Markey, E. (1989) NRC Frowns on Producing Tritium for Weapons in Commercial Plants. *Nuclear Fuel*, May, 13–14.

Martin, E.B.M. (1982) Health Physics Aspects of The Use of Tritium. *Occupational Hygiene Monograph*, No. 6, Leeds.

Mello, G. (1997) New bomb, no mission. *The Bulletin of the Atomic Scientists*, May/June, 20–32.

Moghissi, A.A. and Carter, M.W. (eds) (1973) Tritium. In *Conf. Proc.*, August 30 – September 2, 1971, Las Vegas. Nevada.

Morland, H. (1980) Tritium: Secret Ingredient of the H-bomb. *The Nation*, July 12.

Müller, H., Dembinski M., Kelle A. and Schaper A. (1994) From Black Sheep to White Angel? The new German Export Control Policy. PRIF Reports No. 32, Peace Research Institute Frankfurt, Frankfurt.

Müller, W.D. and Hossner, R. (eds) (1990) *Jahrbuch für Atomwirtschaft 1990*, Düsseldorf.

Müller, E. and Neuneck, G. (1991) *Rüstungsmodernisierung und Rüstungskontrolle.* Baden Baden.

Nuclear Control Institute (NCI) and American Academy of Arts and Sciences (AAAS) (1988) The Tritium Factor. Tritium's Impact on Nuclear Arms Reduction, Washington/Cambridge.

Nuclear Control Institute (NCI) (1989) Future U.S. Production of Nuclear Weapons Materials, April 10, Washington.

Nuclear Energy Agency (OECD) (1988) *The Regulations of Nuclear Trade.* Vol. II, National Regulations, Paris.

Norris, R.S. (1993) Private communication.

Norris, R.S. and Arkin, W.M. (1994) Nuclear Notebook. U.S. Nuclear Weapons Stockpile. *The Bulletin of the Atomic Scientists*, 49, July/August, 61–63.

Norris, R.S. and Arkin, W.M. (1995) Nuclear Notebook. U.S. Strategic Nuclear Forces, end of 1994. *The Bulletin of the Atomic Scientists* ,49, January/February, 69–71.

Norris, R.S. and Arkin, W.M. (1998) Nuclear Notebook. U.S. Strategic Nuclear Forces, end of 1997. *The Bulletin of the Atomic Scientists*, January/February, 70–72.

Norris, R.S. and Arkin, W.M. (1998) Nuclear Notebook. Russian strategic nuclearforces, end of 1997. *The Bulletin of the Atomic Scientists*, March/April, 70–71.

Nonproliferation Treaty (1990) Report of Main Committee II of the 4th NPT Review Conference. Document NPT/CONF.IV/MC.II/1, September 10.

Nuclear Suppliers Group (NSG) (1992) Guidelines for Transfers of Nuclear-Related Dual-Use Equipment, Material and Related Technology, April, Warsaw.

Office of Technology Assessment, U.S. Congress (1987) Starpower. The U.S. and the International Quest for Fusion Energy. OTA-E-338, Washington.

Olgaard, P.L. (1971) Tritium Safeguards. In *Impact of New Technologies on the Arms Race*, B.T. Feld, *et al.* (eds), Proc. of the 10th Pugwash Symposium, Wingspread, Racine, Wisconsin, June 26–29, 1970, Cambridge.

Ontario Hydro (1989) Press release, August 30.

Phillips, J.E. and Easterly, C.E. (1980) Sources of Tritium. Oak Ridge National Laboratory, ORNL/TM-6402.

Röther, W. (1980) Produktion und Freisetzung von Tritium und C–14. In *Strahlenschutzprobleme im Zusammenhang mit der Verwendung von Tritium und C-14 und ihren Verbindungen*, F.-E. Stieve, G. Kistner (eds), STH-Berichte 12/80, Berlin.

Schaper, A. (1991) Arms Control at the Stage of Research and Developement —The Case of Inertial Confinement Fusion. *Science and Global Security*, 2, 279–299.

Schaper, A. (1997) A treaty on the cutoff of fissile material for nuclearweapons —What to cover? How to verify? Peace Research Institute Frankfurt, PRIF Report No. 48, July, Frankfurt.

Scheffran, J. (1989) Strategic Defence, Disarmament and Stability—Modeling Arms Race Phenomena with Security and Costs under Political and Technological Uncertainties. Dissertation in Physics, Marburg and Darmstadt.

Schroeder, Bundesministerium für Verteidigung (1992) Personal Communication.

Seifritz, W. (1984) *Nukleare Sprengkörper—Bedrohung oder Energieversorgung für die Menschheit?* München.

Sinden, D.B. (1986) Tritium related exports. Notice 86–5 by the Atomic Energy Control Board, March 14, Ottawa.

Spratt, P., *et al.* (1985) Final Report of the Tritium Issue Working Group (Vols 1 and 2). September, Toronto.

Stern, W.M. (1988) Nuclear Weapons Material Control: Verification of Tritium Production Limitations. Master thesis, MIT, Cambridge.

Stieve, F.-E. and Kistner, G. (eds) (1980) Strahlenschutzprobleme im Zusammenhang mit der Verwendung von Tritium und C-14 und ihren Verbindungen. STH-Berichte 12/80, Berlin.

Straume, T. (ed.) (1993) Tritium dosimetry, health risks, and environmental fate. Special issue of *Health Physics*, 65, 593–772.

Revealed: The secrets of Israel's nuclear arsenal. *The Sunday Times*, 1986, October 5, and October 12.

Sutcliff, W.G. (1988) Limits on Nuclear Materials for Arms Reduction – Complexities and Uncertainties. *Science*, 241, 1166.

Taylor, T.B. (1989) Verified Elimination of Nuclear Warheads. *Science and Global Security*, 1, 1.

Trutnev, Y., Andrjushin, I. and Chernyshev, A. (1991) On the elimination of the global nuclear threat. *Pugwash Newsletter*, 29, July, 23–25.

Tsipis, K., *et al.* (1986) *Arms Control Verification.*

United Nations Scientific Committee on the Effects of Atomic Radiation (UNSCEAR) (1977) Sources and Effects of Ionizing Radiation.

United Nations Special Committee (1991) Report of the 7th UNSCOM inspection in Iraq. Vienna.

United States National Committee on Radiation Protection (1976) Tritium measurement technologies. NCRP Report 47, Washington.

United States Nuclear Regulatory Commission (1989) Policy Issue (Information) on NRC Export Licensing activities in 1988. SECY-89-080, March 6.

U.S. Department of Energy (1996) Stockpile Stewartship and Management Plan (U). Deleted version, February, Washington.

U.S. General Accounting Office (1991) Controls Over the Commercial Sale and Export of Tritium Can Be Improved. GAO/RCED-91-90, March.

Wall, W.R. and Cruz, S.L. (1985) Tritium Control and Accountability Instructions. SANDIA Report SAND 85-8227.

Wilkie, T. (1984) Old Age can kill the bomb. *New Scientist*, February.

Wilson, E.J. (1954) Manipulation of radioactive gases in high vacuum apparatus. *Vacuum*, 4, 303.

Winterberg, F. (1981) *The physical principles of thermonuclear explosive devices*. New York.

Zuercher, R.R. (1991) NRC, State may seek written assurance targeting retransfer of U.S. tritium. *Nuclear Fuel*, August, 16.

Chapter 2

Diversion path analysis

2.1 Introduction

In nature, tritium occurs at concentrations far too low to make its extraction practically achievable. One reason for this is its comparatively rapid radioactive decay with a half-life of 12.3 years. Since there are no exploitable natural resources, tritium has to be produced artificially by a nuclear reaction. Significant quantities can only be achieved by neutron capture and with a high neutron flux as can be found in nuclear reactors.

Therefore, a diversion path analysis has to take into account possibilities of artificial tritium production as well as opportunities to remove tritium from existing stockpiles. Besides actually performed productions, the potential production capacity is of interest for a diversion path analysis.

A classification of eight facility types which are relevant for tritium controls is presented and the possible paths of tritium from production to disposal are broken down into principal steps at facilities of these types. In order to get a comprehensive picture of tritium in the world, the total number of relevant facilities and their production capabilities as well as current and future inventories are described and summarized in this chapter. Detailed information by country is given in Appendix A. About 50 paths to divert tritium from these sources and facilities are identified.

The diversion path analysis is the first step in developing verification procedures. Only diversion paths with more than one significant quantity per inspection period (here conservatively assumed to be 1 gram per year) are considered with regard to the verification tasks described in Chapter 3.

2.2 Diversion path analysis as a method to derive control tasks

2.2.1 Methodology of diversion path analysis

Diversion means the *clandestine production* of tritium or its *illegal removal* from existing stocks of tritium in any chemical and physical form for unknown and possibly nuclear weapons purposes in a way which is aimed at remaining undetected by any control procedures in place. *Existing stocks* encompass inadvertently generated tritium still enclosed in a matrix as well as pure tritium gas deliberately produced for use.

A *diversion path* is a diversion strategy (see Section 2.2.2) and a related series of technical steps taken to make tritium available for illegal use. The steps necessary for the four principal technical alternatives to produce tritium are shown in Figure 2.3. In general, a diversion path is made up of more than one step. The *diversion path analysis* concentrates on the critical step which constitutes the bottleneck for diversion. Safeguarding efforts concentrate on these steps and include additional measures to improve the control efficiency. As to production, the bottleneck is the nuclear reaction that generates tritium. As to removal, the critical step is the activity to get physical access to a significant quantity of tritium.

In addition to these critical steps, there may be *significant modifications* of materials containing tritium which are relevant steps in a diversion path. These significant steps can be summarized with the term *extraction*. For example, tritium can be extracted from tritiated heavy water or can be enriched beyond an agreed limit. The diversion is complete when the tritium is removed before or after this modification and therefore covered by the diversion path analysis. Controls can be made more effective by including significant modifications in the control procedures in addition to the critical steps of production and removal. If a modification makes tritium suitable for nuclear weapons purposes — while possibly not making much sense for civilian uses — it can be defined as violation of a tritium control agreement even if no tritium is illegally produced or removed. These significant diversion steps (basically extraction) are therefore included in this diversion path analysis.

Tritium can be produced with different degrees of determination. It is produced inadvertently as a by-product of the operation of nuclear reactors, it can be produced deliberately without affecting the normal operation of the facility used, and a reactor can be designed and operated as a dedicated tritium production facility. Besides the actual production figures, the potential production capacity is also of interest for a diversion path analysis.

2.2.2 Diversion strategies

Diversion strategies are a combination of ideas and specific measures taken to avoid or to cheat control procedures. In principle, they can be analyzed only when control procedures have been established. In this monograph, an iterative process is followed. In anticipation of plausible control procedures as outlined in the next chapter, all relevant diversion strategies are considered in the diversion paths as listed in this chapter. When two diversion strategies are similar to each other and the related critical technical steps identical, they are summarized in one diversion path.

There are two classes of diversion strategies:

1. One is to *avoid* control procedures and to act clandestinely. This strategy can be divided into two substrategies:

The facility as such is kept secret or inspection of the facility is not accepted.

Limited inspections are permitted in the facility as long as they bear no risk of uncovering tritium production activities, while these activities are restricted to those which will not affect inspected plant characteristics.

2. The second is to accept inspections but to *cheat* safeguard measures. The second strategy can be divided into two substrategies as well:

   One is to divert or additionally process a quantity of material hoping that it will not be detected by material accountancy because the quantity is small enough to fall within the range of uncertainty.

   The second is to forge the material balance and to divert or process in excess a quantity of material which is equal to the difference between real and reported values. This can be done either by altering the measured figures or by manipulating the measurement instruments.

2.2.3 Safeguards development methodology

The formal procedure to derive control tasks from a diversion path analysis can be seen from Figure 2.1.

For each path the maximum diversion rate R_{max} is estimated (worst case scenario). Only diversion paths by which more than one significant quantity SQ per warranted detection time t_d can be acquired are considered in the safeguards concept. In this monograph the conservative assumption SQ $= 1$ g and $t_d = 1$ year are made.[1] Termination and exemptions of safeguards are defined accordingly. Whenever tritium is not retrievable at a rate exceeding 1 g/y, it can be excluded from safeguards (see Figure 2.1). Therefore, in this chapter the conditions to produce, remove, or extract more than 1 g/y are described for all paths.

For all appropriate diversion paths, safeguard procedures have to be developed which cover this particular path (see Section 3). If the suggested safeguard procedures were not effective in detecting the given diversion path with the desired probability, they would have to be improved. If that were not possible, this particular path would render a tritium control system incomplete or unfeasible (see Figure 2.1).

2.3 Facility types and flow paths with relevance to tritium diversion

As far as facilities are concerned, safeguards have to be applied whenever either the inventory, the annual throughput, or the maximum potential annual production of tritium exceeds one gram.

A careful analysis of all relevant facilities reveals that it is possible to define eight categories of facilities by the way in which tritium occurs and how it is typically treated.[2] Each of these facility types requires control procedures specifically designed for this type (see Section 3.7).

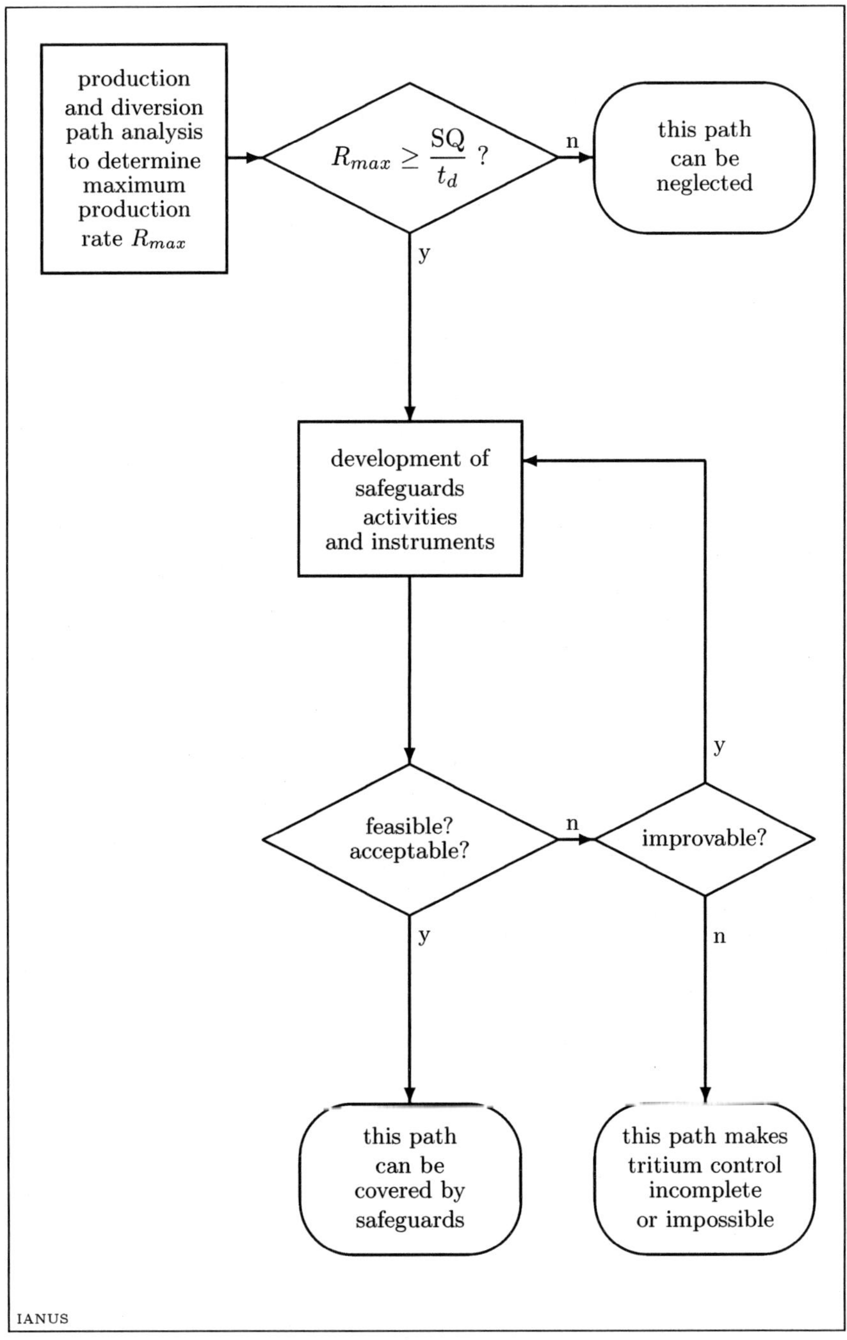

Figure 2.1 Safeguards development methodology.

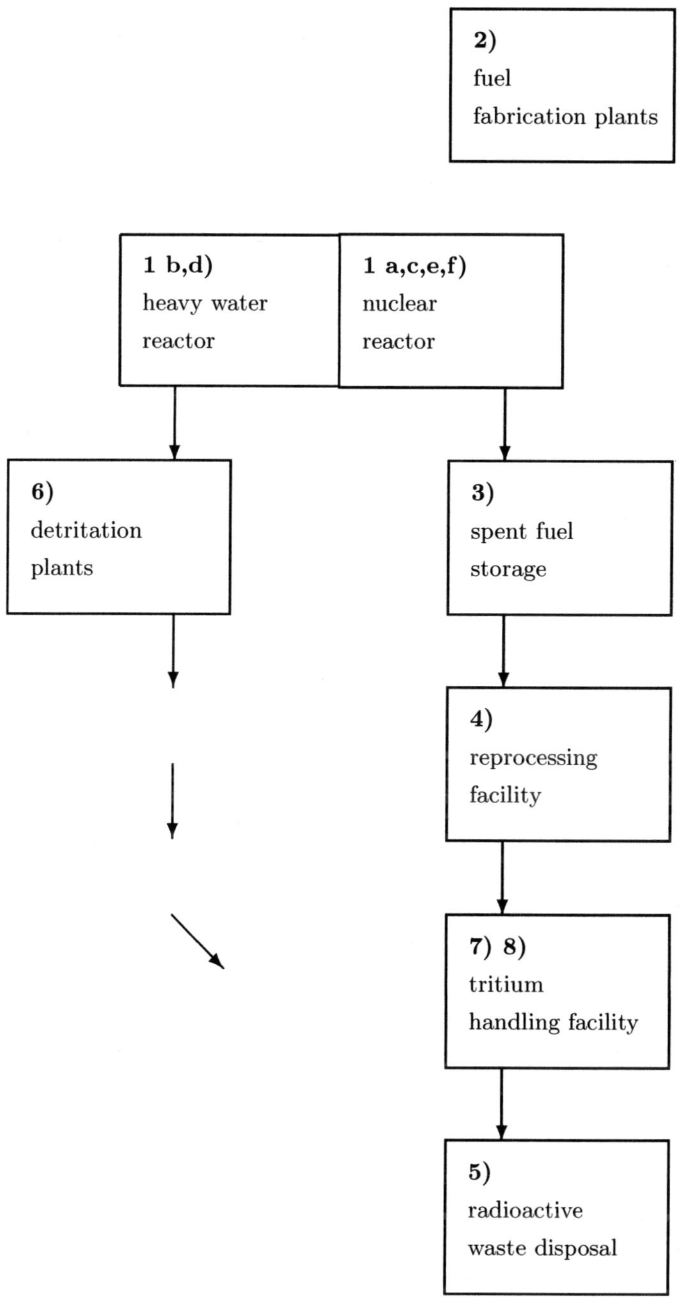

Figure 2.2 Typical flow of tritium through different facility types.

Figure 2.2 shows the typical flow of tritium through the different types of facilities. The numbers given in this diagram refer to the numbering used for facility types in Section 2.6, Appendix A, and Table 3.4. The upper part shows the facilities (type 1) in which tritium is produced either deliberately or inadvertently. The lower part shows the facilities (types 3 to 8) in which tritium is handled and can be removed. This diagram shows only facilities which are of significance to tritium control. Other types may be added if appropriate. For example, if a civilian production line using the lithium-6 path were established and able to generate more than one gram of tritium per year, the facility in which the tritium is removed from the targets by heat treatment would be included in the diagram as a third track parallel to the boxes with the numbers 3) and 6).

2.4 Production paths

Figure 2.3 shows the four *principal production paths* for tritium. It should be noted that the *typical flows* through various facility types in Figure 2.2 are relevant but not identical to these principal production paths. The *principal production paths* are those technical paths which can yield a significant production rate. They are named after the required raw materials and are differentiated in several alternative production paths based on the same raw material. A fifth raw material for tritium breeding is boron, but it is not included in Figure 2.3 because it has a comparatively low maximum production rate. Altogether, 55 different *diversion paths* are described in this monograph. These are 43 production and 12 *removal paths.*[3]

In Figure 2.3, the principal production paths are broken down into various steps from preparation to production and further to disposal. The critical steps are carried out at facilities of the eight types. The numbers in brackets refer to the numbering of facility types which are relevant for tritium control as given in Table 3.4 and in Section 2.6. This diagram contains a simplified model and shows the main relations. Additional connection lines could be added, e.g., from intermediate steps directly to waste storage. The critical steps are described in this chapter.[4]

The upper part shows the four main production paths. These are covered in this section. In the part below the "neutron flux" section, each process step represented by a box and each transfer between facilities offers opportunities for illegal removal, encounters losses to the environment as well as radioactive decay, causes hold-ups, and generates tritiated waste. This is dealt with in the sections on removal paths (Section 2.5) and material accountancy (Section 3.4).

2.4.1 Lithium-6 path

Lithium occurs naturally in two stable isotopes: lithium-6 (7.5%) and lithium-7 (92.5%). Both isotopes undergo nuclear reactions induced by neutrons which lead to the formation of tritium. The reaction with lithium-6 has a very large cross-section at thermal neutron energy,[5]

$$^{6}\mathrm{Li} + n \rightarrow \alpha + {}^{3}\mathrm{H} + 4.8\ \mathrm{MeV}\,, \tag{2.1}$$

the reaction with lithium-7 works only with fast neutrons.[6]

$$^{7}\mathrm{Li} + n \rightarrow n + \alpha + {}^{3}\mathrm{H} - 2.5\ \mathrm{MeV}\,. \tag{2.2}$$

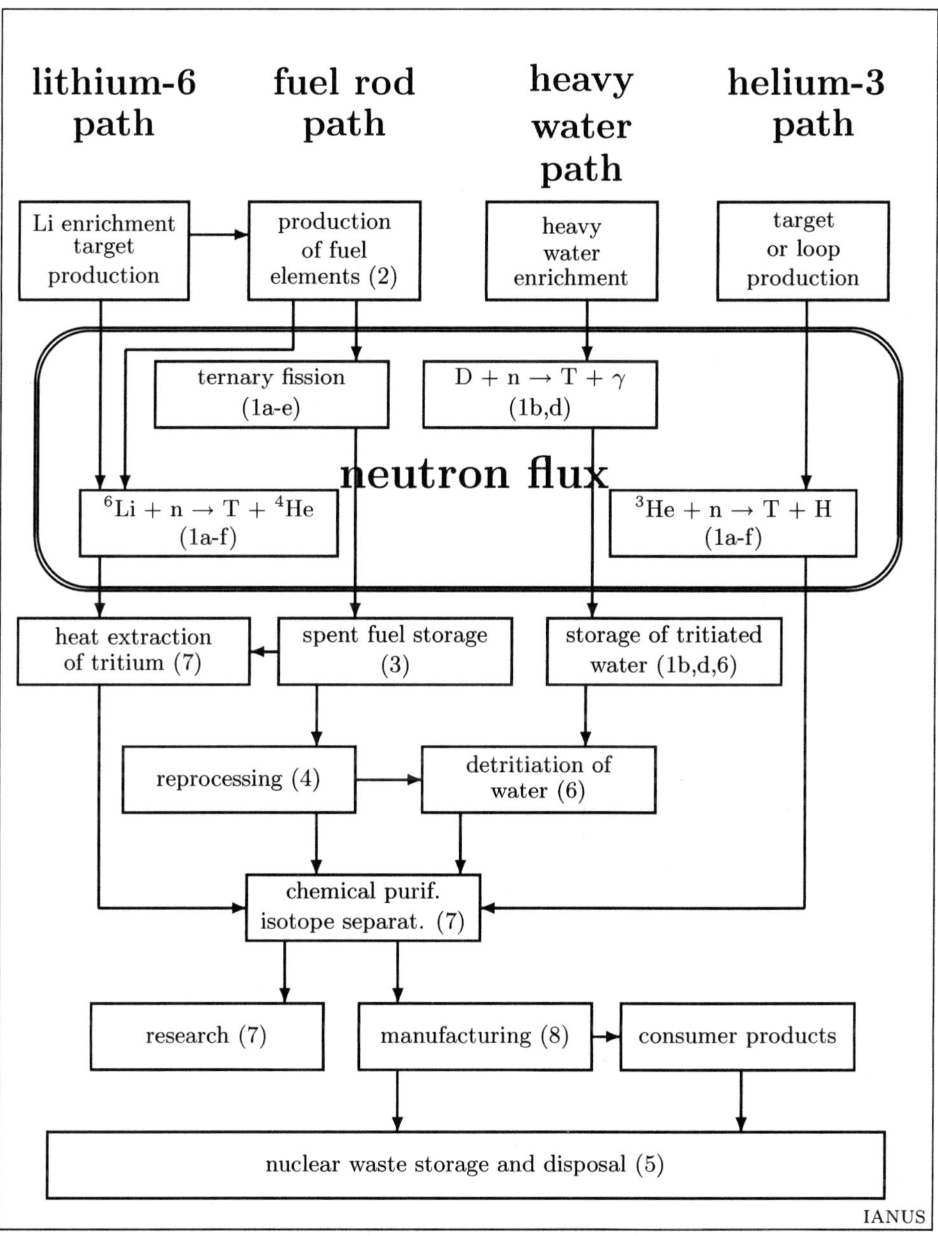

Figure 2.3 Steps of the four principal tritium production paths in various facility types from preparation to disposal.

The production of ^{6}Li from ^{7}Li and subsequent production of tritium via the first reaction can be neglected because the cross-section is low and the threshold energy high.[7]

$$^7\text{Li} + n \rightarrow 2n + {}^6\text{Li}. \tag{2.3}$$

Prior to irradiation of targets containing lithium-6 in a nuclear reactor, the lithium ore has to be mined and milled, converted to LiCl, and transformed into metal by electrolysis (see Figure 2.3). The enrichment in lithium-6 is most commonly done by a process which uses large amounts of mercury.[8] Targets are produced as aluminium alloy or ceramic material. After irradiation the tritium is extracted in a vacuum oven, purified chemically, and separated from the other hydrogen isotopes. Tritium requires special handling and storage facilities due to its radioactivity and because it is a very mobile gas.

Different varieties of this path can be distinguished by the specific mode of operating the reactor and inserting targets into its core (see list at the end of this subsection).

The rate of tritium production by breeding from lithium-6 (or any other raw material) depends on reactor characteristics as well as on the target design and location. This can be expressed with the *specific production rate* F_{rate}, which can be defined empirically by the following production rate equation:

$$m' = F_{rate} \times P_{th} \times t_{exp} \times F_{cap}(t_1, t_2) \tag{2.4}$$

where

m' – production rate

F_{rate} – specific production rate, the most important and usually empirically determined factor in this equation[9]

P_{th} – thermal power of the reactor, which is roughly proportional to the neutron flux in and around the reactor core

$t_{exp} = t_2 - t_1$ – exposure time beginning at time t_1 and ending at time t_2

$F_{cap}(t_1, t_2)$ – capacity factor during exposure time

The maximum production rate can be achieved with full capacity ($F_{cap}(t_1, t_2) = 100\%$). The actual amounts produced depend on operating power levels and the period of time the reactor is running.

For each diversion path discussed in the following list a specific production rate can be given, measured in produced mass per time and power output. Specific production rates depend heavily on the reactor type and mode of operation. Examples are summarized in Tables 2.1 and 2.2.

It has to be pointed out that tritium production typically competes with plutonium production and may be compromised by electrical power production. If nothing else is stated, the specific production rate expresses the case of the reactor being dedicated entirely to tritium production.

Table 2.1 Tritium production via lithium path in power reactors and their specific production rate related to electrical power, shown for certain reactors.

reference	facility	production rate [g/y]	specific production rate [TBq/(MW_ey)]
	with significantly reduced power production		
Donnelly (1989)	any 1 GW_e nuclear power plant, capacity factor 70%	1800	650
	without affecting normal operation		
Ragheb (1981)	1 GW_e LMFBR, lithium coolant instead of sodium	3330	1200
Ragheb (1981)	1 GW_e LWR, fuel modified to produce tritium instead of plutonium-239, fuel enriched to 3%, 65% cap. factor	6510	2340
Ragheb (1981)	1 GW_e CANDU, fuel enriched to 2%, 65% cap. factor	19500	7020
CFFTP (1988)	1 GW_e LWR, at costs of 2–3 times the current commercial market price	100	36
Dastur and CFFTP/AECL (1986)	1 GW_e CANDU, at costs of roughly twice the current commercial price	200	72
Lu *et al.* (1988)	1 GW_e PWR, 600–900 target rods in free control rod guide tubes	100	36
Ragheb (1981)	1 GW_e PWR, eight instrument wells filled with solid lithium pellets	100	36
Ragheb (1981)	1 GW_e PWR, burnable poison rods containing ^{6}Li instead of B	80	29
Benedict *et al.* (1981)	1 GW_e PWR, burnable poison rods containing ^{6}Li instead of B	6	2.2

Table 2.2 Tritium production via lithium path in research or dedicated production reactors and their specific production rate related to thermal power, shown for certain reactors.

country	facility	thermal flux $[s^{-1}cm^{-2}]$	production rate [g/y]	specific production rate $[TBq/(MW_{th}\,y)]$
	at dedicated production reactors			
	theoretical limit (each neutron produces one tritium atom)			4400
	practical limit estimated for tritium safeguards: 200 kW_{th}		1	1800
France	Celestin I, II, 250 MW_{th} heavy water reactors		600 [a]	865
India	Dhruva (BARC), 100 MW_{th} heavy water research reactor [b]	1.8×10^{14}	300 [c]	1080
Israel	IRR-2, 150 MW_{th} heavy water research reactor at Dimona		500 [d]	1200
Russia	HWR with natural uranium, 50 MW_{th}		55 [e]	400
	graphite pile, 250 MW_{th}		18 [f]	26
U.S.	SRP (Savannah River Plant); 2400 MW_{th} HWR, fuel enriched to 93%		6300–11500	950–1730 [g]
	SRP; 2400 MW_{th}; control rods		500–900	70–130 [h]
	SRP; 2400 MW_{th}; in addition to Pu production		500–900	70–130 [i]
	Fast Flux Test Reactor at Hanford (DOE), 400 MW_{th}		1200 [k]	1000
	Advanced Test Reactor at Idaho (DOE), 250 MW_{th}	8.5×10^{14}	700	1000
	High Flux Isotope Reactor at Oak Ridge (DOE), 100 MW_{th}	21×10^{14}	300	1000
	High Flux Beam heavy water reactor at Brookhaven National Laboratory, 40 MW_{th}	7×10^{14}	20 [l]	180
	Oak Ridge Research Reactor, 30 MW_{th}		30	360
	Materials test reactor, 10 MW_{th}		15–30	540–1080
	All three in-core sample assemblies of the 5 MW_{th} MITR-II research reactor replaced by lithium targets	0.8×10^{14}	1.5	108

Table 2.2: *(continued).*

country	facility	thermal flux [$s^{-1}cm^{-2}$]	production rate [g/y]	specific production rate [TBq/(MW_{th} y)]
	Based on continuously running 3000 MW_{th} of the U.S. naval reactors		≈10,000 [m]	1200
	Lithium-cooled fast reactor, 2700 MW_{th}, 70% cap. factor		≈25,000 [n]	4800
	with simultaneous power production			
	N-reactor at Hanford, 4000 MW_{th}, fuel enriched to 2.1%		3000 [o]	270
	N-reactor at Hanford, 4800 MW_{th}, fuel enriched to 2.1%		6250	470 [p] (+ 815 g Pu)
	N-reactor at Hanford, 4800 MW_{th}, higher enrichment		16480	1240

[a] See Gsponer (1984). According to Gsponer (1984) the production may even be as high as 1 to 2 kg/y in a dedicated mode of reactor operation.

[b] Dhruva was completed in August 1985 and reached full power in early 1988. It is not clear whether this unsafeguarded reactor is used as a production reactor. The given figure is the potential production capacity.

[c] Maximum production rate without simultaneous Pu production if running at full capacity and totally dedicated to this purpose.

[d] same as *c*

[e] 100% capacity is assumed. Data are taken from a CIA study which was conducted in the mid-50s and quoted in Cochran (1989).

[f] 100% capacity is assumed (Cochran, 1989).

[g] The lower limit applies for (realistic) 200 days of exposure, the upper limit applies for a capacity factor of 100%. Rule of thumb for the specific production rate: 1 g ^{235}U produces 0.95MW_{th}d and 0.9/72 g = 1/80 g of tritium. 0.9 g/(MW_{th}d) (Cochran, 1987a) is the plutonium equivalent production and 72 is the weight ratio of tritium and plutonium. About 40% of the neutrons are consumed for tritium production. This specific production rate assumes no production in control rods.

[h] Production in control rods in addition to dedicated tritium or plutonium production (Cochran, 1987a).

[i] Production in blankets and dischargeable targets inside driver assemblies in addition to plutonium production (Cochran, 1987a).

[k] This and the next two examples are given in Donnelly (1989). The estimation was based on 200 days' exposure time and on a specific production rate of F_{rate} = 14 mg/MW_{th}d for reactors fueled with highly enriched uranium or plutonium.

[l] This and the next three examples are given in Stern (1988).

[m] See Morrison and Tsipis (1988). This is a very rough first estimate.

[n] See DeVolpi (1987). This is a very rough first estimate, probably far too high.

[o] See Cochran (1987b).

[p] See Ragheb (1981).

Calculation of the production rate. The specific production rate can be derived empirically from operational experience. It provides a rough estimate of production capacities. For a more thorough calculation, the local production rate must be known. An approximation for the local tritium production rate $R_{\alpha,i}$ is given by

$$R_{\alpha,i} = \Phi_i \times \sigma_\alpha \times \rho_{Li-6,i} \quad (2.5)$$

where

$R_{\alpha,i}$ – Local reaction rate of nuclear reaction of type α, e.g., ^{6}Li(n,α)^{3}H, and at point i, i.e., target location in the reactor core. $[R_{\alpha,i}] = \mathrm{m}^{-3}\mathrm{s}^{-1}$

σ_α – Reaction cross-section of reaction α. $[\sigma_\alpha] = \mathrm{cm}^2$

Φ_i – Neutron flux at point i. $[\Phi_i] = \mathrm{cm}^{-2}\mathrm{s}^{-1}$

$\rho_{Li-6,i}$ – Density of lithium-6 at point i. $[\rho_{Li-6,i}] = \mathrm{m}^{-3}$

The overall production rate via lithium path P_α of one facility can be calculated by summation (or if appropriate by integration) over the different target locations:

$$P_\alpha = \sum_{all\ i} (R_{\alpha,i} \times V_i \times t_i) \frac{M_T}{N_{Av}} \quad (2.6)$$

where

P_α – Production rate. $[P_\alpha] = \mathrm{g/y}$

V_i – Target volume. $[V_i] = \mathrm{m}^3$

t_i – Exposure time of target at location i per year of reactor operation. $[t_i] = \mathrm{s/y}$

M_T – Molecular weight of tritium. $[M_T] = 3.016\ \mathrm{g\ mol}^{-1}$

N_{Av} – Avogadro constant. $[N_{Av}] = 6.02 \times 10^{23}\ \mathrm{mol}^{-1}$

i – Possible target locations in nuclear reactors and irradiation space, which can be obtained by replacing other materials.

The microscopic production rate of tritium is proportional to two factors: (1) the neutron flux, and (2) cross-section of the breeding reaction, i.e., the probability of neutron capture in the target. Accurate accountancy of tritium depends on precise knowledge of the production rate. It seems not to be feasible, even with considerable effort to calculate the production rate with the required accuracy, because there are too many varying parameters. The production rate would have to be calculated for each single lithium target individually. The neutron flux is not homogeneous over the reactor core. Thus, the production rate depends on the target position. The flux depression in the target, which is due to the high cross-section of lithium-6, causes a spatially varying flux inside the target. Thus, the production rate depends on the target geometry.[10]

The average thermal neutron flux $\langle\Phi\rangle_x$ in a thick target as a function of the target thickness x is calculated by the equation

$$\langle \Phi \rangle_x = \frac{\Phi_0}{x/2} \int_0^{x/2} exp\,(-\rho_{Li-6,i}\,\sigma_\alpha\,s)\;ds\,. \tag{2.7}$$

Measurement of the production rate. In practice the tritium production rate is typically determined either from the ratio of lithium-6 consumed and the initial content of lithium-6 (see, e.g., Hugony *et al.*, 1973) or from direct measurement of extracted tritium rather than by theoretically calculating the neutron irradiation rate of the lithium target.

Production time. The time required to start an irradiation program in an operating nuclear reactor has to be considered (Rupp *et al.*, 1965). It depends on the time required to fabricate suitable targets and whether irradiation space is currently existing in the reactor (e.g., stationary target sites) or changes in the reactor are necessary. Without changing the enrichment level and for simple target fabrication, about one year is required. If a large tritium production program was intended, a complete redesign and test of the new fuel assemblies would be necessary. This can only be achieved with a major effort and would take at least two to three years of work. In addition, a lead time for an investigative program of about one year has to be taken into account. Fuel enrichment becomes necessary in two cases:

1. When fuel is replaced by targets, enrichment would be necessary to compensate for loss of fuel.

2. When the production cross-section σ_α is high (i.e., the production rate is high), enrichment would be necessary to compensate for the added poison. No fuel enrichment is required if the production cross-section is fairly low (i.e., lies in the range of 10 mbarn to 1 barn). In case of high cross-sections ($\sigma_\alpha \simeq 100$ barn), a considerable enrichment of the reactor fuel is required (an increase from 3.8% to 4.3% is given as an example) (Rupp *et al.*, 1965).

In the case of dedicated production reactors held in standby mode or producing a limited quantity of tritium below the maximum production capacity, the production time is much smaller since no preparations of the reactor core are necessary. For example, one reactor at Savannah River Plant (SRP) could produce at maximum 100 g of tritium within three days. This would, however, be highly irrational, since a very small amount of tritium would be distributed in tonnes of target material. Production is made in batches and irradiation continues until about 5 to 10% of the lithium-6 has been transformed. According to the normal production scheme at SRP, a period of some six to eight months is required for target irradiation. This does not, of course, include time for target loading, reactor startup and shutdown, target removal, cooling, extraction, purification, enrichment of tritium, fabrication and transport to the nuclear weapons storage site. If time were pressing, a quick production cycle could probably be completed after three months.

Minimum quantity of lithium-6 to be detected. Due to buckling,[11] lithium-6 targets can be depleted only by 50%. Thus, a minimum of 4 g of lithium-6 is required to produce one significant quantity of tritium (1 g) with one single target. To arrive at this high burnup, the target has to be placed in a reactor for more than one year.

The maximum achievable burnup within one year and with a thermal neutron flux of 10^{13} $s^{-1}cm^{-2}$ would be 20% (see Figure 2.4). Thus, 10 g of lithium-6 are required. From this it follows that the detection limit for lithium-6 has to be 10 g if the verification goal is to detect any targets capable of producing 1 g of tritium.

The maximum density of lithium-6 in the target is about 0.1 g/cm^3. Therefore, the target volume has to be at least 100 cm^3, i.e., less than the volume of a single PWR fuel rod would have to be filled with target material.

Diversion paths based on lithium

1. Unreported tritium breeding in military production reactors which are declared not to be used for tritium production.

   (a) Resume operation of a shutdown production reactor without power generation.

   Fuel/lithium assemblies especially designed for tritium production are used. These assemblies look quite different from conventional fuel assemblies. Production reactors are especially designed and optimized for producing either plutonium or tritium. In the optimum case they are moderated by heavy water, operated at low temperature, and do not produce electrical power. The production reactors at the Savannah River Plant (SRP) were of this type. The experience of the 2400 MW_{th} production reactor at the Savannah River Plant can be used as a rule of thumb. Accordingly, a dedicated reactor (i.e., a reactor designed and optimized for the production of tritium) can produce about 1/72 g of tritium for each gram of uranium-235 fissioned, which is accompanied by the release of one megawatt-day of thermal energy ($MW_{th}d$). Hence, the specific production rate is $F_{rate} = 1/72$ g/($MW_{th}d$) (1/72=0.0139). The actual design-dependent production rates are within ±20% of these values (Cochran, 1987a, p. 59).

   At SRP the lithium charges can either be combined with highly enriched uranium fuel (Mark 16B and 22), or put in a separate target (Mark 60B). From the little information not held as classified, it can be derived that each Mark 22 assembly contains roughly 70 g of lithium-6 and produces between 13.9 and 17.4 g of tritium.[12]

   For other examples, see Table 2.2.

   (b) Resume or start up tritium production in a production reactor which is operated for power production.

   Russia still has three reactors in operation which produce plutonium as well as electricity and heat (see Table 2.6). Their production capacity for tritium is not known, but estimated to be 2 kg for each reactor.

The N-R reactor at Hanford (U.S.), which was shut down in 1987, can serve as an example. It is graphite-moderated and is cooled with pressurized light water. It was mainly operated to produce plutonium and simultaneously produced 862 MW_e electricity. Although this reactor is a more effective producer of plutonium, the production of tritium has been demonstrated. 650 kg/y weapon-grade plutonium and 3 kg/y tritium can be coproduced (Cochran, 1987a).

For other examples, see Tables 2.1 and 2.2.

2. Convert power or research reactor to a dedicated production reactor without or with remarkably reduced simultaneous power production.

   (a) Especially prepared fuel composition which contains lithium-6 instead of uranium-238 with no changes in core and assembly geometry.

   If all neutrons otherwise consumed for breeding plutonium-239 from uranium-238 were to breed tritium from lithium-6, 6.5 kg could be produced per GW_ey (Ragheb, 1981). The assumptions made are fuel enrichment of 3%, burnup of 30,000 MW_ed, and a capacity factor of 65%. This estimation gives the theoretical maximum for tritium production and can be achieved only to a limited degree, since this replacement of uranium-238 by lithium-6 with its high neutron absorption cross-section could not allow criticality if distributed over the core.

   (b) Especially designed fuel assemblies used in the whole core.

   The HTGR especially has been considered by the U.S. as a new tritium production reactor, because it is well suited both as tritium production reactor and as power reactor. But a different core design is required for the two applications (Nuclear Control Institute and The American Academy of Arts and Sciences, 1989, p. 66).

   A U.S. patent has been granted for an LWR fuel assembly for the production of tritium (Cawley and Trapp, 1985). This assembly looks markedly different from conventional fuel assemblies. It consists of two intermashing arrays of subassemblies. The array of the first subassemblies is 3 × 3, and the array of the second assemblies is 2 × 2.

   For example, in CANDU reactors, enriched and special production assemblies could be used in the whole core. If the fuel were enriched to 2% and the reactor ran at 65% capacity factor, a production of 19.5 kg/(GW_ey) is believed to be possible (Ragheb, 1981). However, this estimate seems to be too high.

   The specific production rates for various reactors and production schemes given in Tables 2.1 and 2.2 can be summarized as ranging from 1000 to 5000 g/(GW_{th}).

3. Unreported tritium breeding in power and research reactors without significant changes of the core and fuel assemblies and without significantly affecting normal operation.

   Large quantities of tritium can be produced in any nuclear reactor by altering core design, fuel enrichment, and/or moderator. These cases are dealt with in the previous path. Smaller amounts can be produced without any changes in these characteristics. For examples, see Table 2.1.

The primary purpose of power reactors is to generate electricity. In estimating the tritium production capacity of power reactors via the lithium path only such production schemes are considered which would not adversely affect the reactor operation, i.e., which would result neither in a noticeable decrease of the electrical capacity nor in changes of reactor characteristics (e.g., power peaking, reactivity coefficient) that significantly affect reactor safety. In any case, a significant development effort is required that would take some years' time.

These assumptions are made in order to define a class of production scenarios which pose the most demanding requirements to inspection activities. They are characterized by not being detectable by simply observing operational parameters of ordinary power production (reactivity holddown, axial and radial power distribution, cycle length). Estimations of the tritium production capacity have to consider that the measurement accuracy of these parameters is about ±5% (Lu *et al.*, 1988). In this study, it is not assumed that the tritium production would make sense economically. Estimates of tritium production by LWRs are about 0.1 kg/(GW_ey), at costs of 2–3 times the current commercial market price (CFFTP, 1988). CANDU reactors could produce about 0.2 kg/(GW_ey) at costs of about two times the market price or even less, if enriched uranium fuel were used (Dastur and CFFTP/AECL, 1986). However, if the costs are not an issue, a higher "worst-case" production capacity can be achieved.

The radio-isotope production capacity in a reactor cannot be accurately determined without a careful engineering study of the particular reactor system. Details to be considered are as follows:

- Build-up of gas pressure inside the target restricts the quantity of tritium that can be produced. The target has to be removed before capsule failure occurs. If no direct access to target positions is available during operation, it should be assumed that the insertion and removal of the targets would coincide with the normal refuelling cycle of the reactor (6–18 months, depending on the reactor). Target changes more frequently than shutdown can be performed only if special facilities are available which, for example, penetrate the reactor pressure vessel.
- Typically well-thermalized neutron fluxes of $1–3\times10^{13}$ cm^{-2}s^{-1} are available in power reactors. Thermal fluxes below 1×10^{13} cm^{-2}s^{-1} are considered unsuitable if significant production rates are required (Rupp *et al.*, 1965).
- Many power reactors are operating at fairly high temperatures and pressures.
- Most boiling water reactors and some pressurized water reactors have completely filled fuel assemblies or bundles, i.e., all positions are filled by fuel rods (Rupp *et al.*, 1965). Hence, in order to make available target space, fuel would have to be displaced.

- Further inconveniences of lithium targets are the addition of poison and the reduction of the heat-producing surface.

The following production paths can be distinguished:

(a) Replace conventional fuel rods with rods containing lithium-6.

Lithium-containing materials considered for this use are Li-Al alloy,[13] lithium silicates, aluminates ($LiAlO_2$), and aluminosilicates. All these target materials have the disadvantage that the build-up of gas pressure inside the target capsule requires early removal of the target or restricts the volume-specific production rate. A new alternative is a target of lithium aluminate particles dispersed in a matrix of zirconium.[14] The gas pressure within the capsules containing the target material can be reduced by zirconium in which tritium is readily absorbed.

As an estimate, 2 g of tritium can be bred within one year if only a single fuel rod in a PWR fuel assembly is replaced by a rod filled with lithium-6 at the maximum possible density. For this estimation, an active target length of 300 cm, an inner rod diameter of .95 cm (i.e., the volume is 210 cm^3), and a maximum concentration of 0.1 g cm^{-3} (0.067 g cm^{-1}) for lithium-6 are assumed. The lithium-6 content of the target rod would be 21 g. Therefore, a burnup of 10% would suffice to breed one gram of tritium. The maximum achievable burnup of 20% (see above) would yield 2 g.

Such a high yield per rod cannot be achieved without affecting normal operation if several hundred target rods were in the reactor core at the same time. A maximum of 100 g/(GW_ey) tritium can be bred (CFFTP, 1988).

Even partly lithium-6 fillings of single fuel rods should be detectable.

(b) Replace conventional control rods by those made of lithium-6.

The material considered for this purpose is Li-Al alloy (3 wt% Li, 40% enriched in lithium-6) (Ragheb, 1981) and lithium aluminate/zirconium.

The specific production rate for such "secondary targets" in a production reactor at the Savannah River Plant dedicated to plutonium or tritium production can be estimated at F_{rate} = 0.001 g/MW_{th}d (Cochran, 1987a). If this is transferable, a 1 GW_e power reactor with 33% efficiency running at 80% capacity would produce about 900 g/y.

In a similar way, external absorber rods could be replaced by lithium bundles, or appropriate fuel management strategies could be applied to provide flux flattening in order to allow all fuel assemblies to operate at similar power levels near their operating limits. Otherwise, the entire reactor would be limited by the few hottest assemblies. In CANDU reactors, about 200 g/(GW_ey) could be produced (Dastur and CFFTP/AECL, 1986).

(c) Replace boron burnable poison rods (BPR) with lithium BPRs (for PWRs only).

In the U.S., BPR which contain $LiAlO_2/Zr$ instead of boron are under development. The production rate is estimated for a 1 GW_e reactor at 80 g/y (Ragheb, 1981) and 6 g/y by another reference.[15]

(d) Insertion of target rods in empty spaces inside the core.

The lithium targets are inserted in guide tubes in place of thimble plugs and burnable poison rods. At the same time, targets are inserted in control guide tubes that are not positioned beneath control rod clusters. Alternatively, lithium-containing rods can be inserted in empty grid spaces or in unused fuel positions in the core, if available.

If 30 to 45 assemblies in the peripheral region of the active core of PWRs are loaded with 20 lithium-containing rods in each assembly, some 100 g of tritium can be produced in one reactor cycle.[16] The specific production rate is 30 g/(GW_ey), but an even higher rate can be achieved by changing more fuel assemblies (about 200 make up the complete core). A total of six to nine rods would be needed to produce 1 g/y.

Alternatively, approximately 60 fuel assemblies in the internal region of the core with eight lithium-containing rods each would serve the same purpose, i.e., a total of five rods would suffice for 1 g/y (see item below). This case involves the substitution of the burnable poison rods by lithium-containing rods. This implies that the safety margin of the reactor is severely reduced, but possibly not so much as to necessarily preclude this diversion path.

For BWRs this scenario would require major changes in fuel assemblies to introduce enough lithium-containing material, since there are only two water rods in an assembly, and one of them is an anchor rod. Therefore, this scenario is not considered for BWRs.

(e) Insertion of lithium targets in empty spaces outside the core (if available).

In particular, the periphery of the reactor, empty regions outside the reactor core, and above and below the active core are considered.

Appropriate positions are not available in all reactors. In particular, new designs of reactor cores are optimized in such a way that no space is left unused. This production path would probably cause the least flux depressions and interferences with normal reactor operation,[17] but the production rate would not be very high because the neutron flux is considerably lower than average values.

A study on unreported plutonium production in LWRs came to the conclusion that this path is not of great safeguards concern, i.e., no more than a significant quantity of plutonium (SQ = 8 kg) could be produced (Lu *et al.*, 1988). This has to be reexamined for tritium because the tritium equivalent to one SQ of plutonium is 100 g of tritium which is much more than the SQ of tritium (i.e., 1 g).

4. Using irradiation target positions in research reactors.

Tritium breeding experiments for fusion research can serve as an example. A facility was constructed at the Japan Material Testing Reactor (JMTR) to produce 40 TBq (0.1 g) of tritium per batch (Tanase, 1988). Lithium-aluminium alloys are used with 3.2 wt% lithium enriched to 95.5% lithium-6. The size of the target is 80 mm×14 mm×2 mm. They are irradiated for 40–70 days at a neutron flux of 2×10^{14} $cm^{-2}s^{-1}$ and cooled for a few years to let activation products like zinc-65 and iron-59 decay. The burnup of lithium-6 was 20 to 40%, and about 20 targets were used in each 40 TBq batch.

In the fusion research experiment CRITIC 100 g LiO_2 was irradiated for 21 months; 1% lithium burnt, yielding some 0.2 g (Miller, 1988).

In an experiment to test a device for continuous tritium removal from a target containing 780 g LiF and exposed to an average neutron flux of 10^{10} $s^{-1}cm^{-2}$ in a self-sustaining neutronic reactor, an output of 0.28 cm^3/d was achieved (Jenks, 1963). Extrapolation shows that given a neutron flux of 10^{13} $s^{-1}cm^{-2}$ is achievable, the production rate could be 1.4 g/y.

In three test series with target slugs (7.6 cm length, 3.3 cm diameter) containing 160 g of a lithium-aluminium alloy (with 3.5 wt% natural lithium) which were irradiated with a neutron flux of 1.6×10^{13} s^{-1}/g(Li), it was shown that the tritium yield is proportional to the irradiation time (Abraham, 1963). The yields were 0.34 TBq (20 days), 2.0 TBq (75 days), and 5.2 TBq (irradiation time not given), respectively. From the first series it would follow that 6.2 TBq/y (17 mg/y) could be produced per slug.

5. Diversion during reported and inspected tritium breeding with lithium targets.

   This set of diversion paths is relevant if a limited production for military (or civilian[18]) purposes is permitted and verified.[19] In addition to variations of the diversion paths mentioned in the former two cases, the following paths are possible in this case:

   (a) Falsified report on irradiation of lithium targets. Irradiating more lithium-6 (e.g., larger lithium-6 enrichment of the target, understate number of target assemblies placed in reactor, more lithium per target) or for a longer period of time or with a higher specific production rate (e.g., higher neutron flux, thinner targets with less flux depression in the bulk of the target) than reported.

   (b) Understate the quantity of extracted tritium and overstate tritium content in waste.

   (c) Diversion of irradiated lithium targets before filling the vacuum oven for extraction of tritium.

   (d) Cover up breeding of tritium by fusion breeding blanket research. The worldwide production of tritium from fusion materials research was estimated in 1988 to be less than 0.5 g per year (CFFTP, 1988). Large quantities were gained in Japan (100 TBq, i.e., $\approx$ 0.28 g) (Tanase, 1988) and in Canada (0.2 g) (Miller, 1988).

   As a rough estimate it is assumed that 10% of the full diversion capacity can be diverted from the reported production rate.

6. Accelerator spallation neutron source.

   No appropriate facility exists at present, but estimates on the production capacity are available.[20] 30g/y can be produced using a pulsed proton

LINAC with 5 mA average current at 1.1 GeV, which produces a neutron flux of 2×10^{17} thermal n/s.[21] Another study estimated a production rate of 10 kg/y with an accelerator having a current of 300 mA at 1 GeV.[22] Advanced designs (200 mA continuous at 2 GeV) can produce some 5 (Miller, 1989) or 6 kg[23] of tritium per year.

The technology required for accelerator breeding is not at hand. The world's largest proton accelerators can generate a maximum current of a few mA. Quite large design efforts and experimental testing would still be required for the construction of the breeding target, tritium extraction process, and radiation protection measures. At least five years of concentrated development effort would be required.

7. New reactor concepts.

   (a) Seed-blanket reactor (SBR) concepts may be used where highly enriched uranium oxide mixed with ZrO_2 as inert material is used as seed and is surrounded by blanket elements containing lithium. This would be similar to light water breeder reactor systems (LWBR) (Ragheb, 1981).

   (b) Lithium-cooled liquid metal fast breeder reactor (LMFBR).

   Instead of sodium, lithium could be used as coolant for LMFBRs. Estimations are made that a 1 GW_e power reactor of this kind could produce 3.3 kg/y (Ragheb, 1981). Another study on a 2700 MW_{th} reactor of this type came up with 25 kg/y (DeVolpi, 1987), but this estimate seems to be too high.

   (c) Experimental and proposed power fusion reactor.

   One of the few large magnetic confinement fusion *experiments* is the TFTR at PPPL (see Table A.12). For the final design of the Lithium Blanket Module (LBM) it was expected that 1.3 MBq (3.7 ng) of tritium would be produced in 900 LiO_2 rods per deuterium–deuterium (DD) fusion run and 37 MBq (0.1 μg) per deuterium–tritium (DT) shot. The numbers of neutrons produced per shot in TFTR are 2.5×10^{17} for DD fusion and 5×10^{17} for DT fusion (Engholm, 1983). Assuming 10^4 shots per year, the production rate can be estimated to be 37 μg/y by DD fusion in a Tokamak fusion experiment and 1.7 mg/y in DT fusion. Thus, current experimental fusion reactors will not produce significant amounts of tritium.

   No experimental fusion facilities with a relevant average neutron flux exist at present, but design studies are under way and estimates on the production capacity are available. Since tritium self-sufficiency will eventually be required, fusion *reactors* which can produce at least as much tritium as they consume would have to be developed. A 1 GW_e reactor would generate and burn about 180 kg tritium per year.

   (d) Fusion/fission hybrid reactors.

   No appropriate facility exists at present, but this reactor type does not add fundamental aspects to this diversion path analysis, because both fusion and fission reactors are covered by other paths. Tritium would be kept inside the fusion reactor because its primary purpose is breeding of fission reactor fuel by use of fusion neutrons (see, e.g., Greenspan and Miley, 1981).

2.4.2 Helium-3 path

Helium-3 produces tritium by neutron capture and proton emission[24]

$$^{3}\text{He} + n \rightarrow p + {}^{3}\text{H}. \tag{2.8}$$

A restricting factor for this production path is the availability of helium-3. The naturally occurring helium-3 resources on earth are highly diluted and not economically exploitable, although the global inventory is about 4×10^6 kg (see Wittenberg *et al.*, 1986). The major source of helium-3 is the decay of tritium, but it is not clear how much of the decay product is retained and how much is vented to the atmosphere. Assuming 100% retention and based on current inventories, (3.7 ± 1.4) kg helium-3 can be gained from the U.S. nuclear arsenal. The Mound Research Corporation has an unclassified inventory of 13.4 kg and sells about 1.34 kg/y at a rate of some $700/g (Wittenberg *et al.*, 1986). From tritium decay 50 g/y at most can be gained at HWRs, i.e., approximately 1 kg/y from all Canadian CANDU reactors. Gas from U.S. natural wells contains traces of helium-3, which amount to some 28 kg held in underground U.S. helium storages and an estimated 187 kg in known natural reserves (Wittenberg *et al.*, 1986). The moon surface soil contains in the order of 10^9 kg helium-3 (Wittenberg *et al.*, 1986).

Diversion paths based on helium-3

1. Tritium produced from the helium-3 content in the coolant of a high temperature gas cooled reactor can be extracted and purified.

   For a helium-3 content of 0.2 ppm, the production rate may range from 55 to 210 TBq/(GW$_e$y) (0.15–0.58 g/(GW$_e$y)) depending on the percentage of helium in the core (4–20%).[25] The production rate can be enlarged significantly by enriching the coolant in helium-3. If the fraction of helium-3 is larger than 0.4 ppm divided by the power in GW$_e$, the annual production rate could exceed 1 g.

2. External addition of helium-3 to the moderator system of a reactor.

   An analysis of the feasibility of this production path was performed for a hypothetical 500–600 MW$_e$ CANDU reactor (Thomas and Bereton, 1985). The results suggest that the frequent addition of helium-3 to the moderator water will result in an enhancement of tritium production inventories, which is highly dependent on the rate at which helium-3 irretrievably escapes to the moderator cover gas. Uniform dispersal of helium-3 throughout the moderator system would be achieved by multiple orifice inlets. About 300 g of helium-3 can be dissolved in a moderator (300 t). The problem is that helium-3 tends to escape to the cover gas before it is converted to tritium. Helium-3 escaped to the cover gas has to be regarded as lost raw material for tritium breeding, since current CANDU designs do not promote the saturation of the moderator with the cover gas. The enhancement factor depends heavily on the quantity of helium-3 available. Assuming the total input of 300 g over 30 years, the calculated enhancement factors after 30 years range from 1.19 to 1.28 depending on the input scheme. Following an annual addition of 100 g of helium-3 over 30 years, the enhancement factor after that time would be 3.77.

3. Collect tritium generated during a rapid power excursion (RPE) experiment at research reactors.

   Rapid power excursion experiments are carried out to determine the structural integrity of fuel elements under transient conditions. These experiments require that a reactor fuel is placed in the high-flux region of a research reactor and surrounded by a material which reduces the neutron flux to a low level by absorbing neutrons. The shielding is then rapidly withdrawn, exposing the fuel to a rapidly rising neutron flux. A neutron absorbing gas like helium-3, withdrawn or inserted at a readily controllable rate, can provide the variable shielding needed to produce a well-characterized flux excursion at the location of the fuel pin. For this purpose a ramp testing facility has to be built. At the end of the experiment, the tritium produced can be collected. From a little under one gram to several grams a year could be produced in the course of routine ramp tests.[26]

4. Target loops with continuously circulating helium-3.

   In the same way as for rapid excursion experiments, equipment can be installed in the reactor which allows helium-3 gas to be inserted in a continuous manner. By doing this, a long exposure time can be achieved. The gas can either be released after a certain period of time to extract tritium or otherwise the gas could be processed in a continuous manner. In the latter case, helium-3 is circulated through an additional system of pipes and pumps from the reactor core to a gas separation system and back. The required technology has not yet been developed and will certainly be complex. In particular, a high pressure would be required to compress helium to a density high enough to achieve a significant production rate. This would strongly influence the reactor criticality. Should the high pressure system fail and helium-3 be released, the reactor would very rapidly go supercritical, and a serious accident could result.[27]

   A study made for CANDU reactors showed that direct activation of helium-3 contained in a closed loop would have a half-life (time required to convert half of the helium-3 inventory) of about 11 days. Thus it would be essentially complete within a few weeks without any significant loss (Thomas and Bereton, 1985).

   A study was made on an experimental reactor loop of the NRX reactor at Chalk River Nuclear Laboratory (Canada) (Osborne, 1979). The neutron flux reaching fuel elements was controlled by introducing helium-3 into a stainless steel coil in the annular space around the fuel. The expanded volume of the helium-3 system is 15 l (2 g). The production rate of tritium was 50 MBq/s when the reactor power and the helium pressure were both maximum (1 MPa). At a capacity factor of 23%, this experiment would produce 1 g/y. A molecular sieve was used to collect tritium.

Seeding of the annulus gas system in CANDU reactors with helium-3 has been suggested as a method to produce tritium (Thomas and Bereton, 1985).

If loops are installed and the helium-3 content exceeds the quantity required to produce one significant quantity of tritium per desired detection time (e.g., 1 g/y), inspections would be necessary.

5. Closed helium-3 targets.

   Helium-3 can be put into a pressure cell. This target type has already been developed for cross-section experiments. One example is a 1.6 mm thick stainless steel cell filled to a pressure of 140 atm (1.379×10^4 Pa). The cell was a 3.4 cm diameter, 3 cm high cylinder with a hemispherical dome and a volume of 40 cm^3 (Ward, 1981). Assuming that the pressure in this target should reach 140 atm only at the maximum fuel temperature T_{max} (Lu *et al.*, 1988), it can be filled with 7.8 kg in case of BWRs (T_{max} = 2160 K) and 6.1 kg in case of PWRs (T_{max} = 2760 K). If an average neutron flux of 10^{12} $s^{-1}cm^{-2}$ can be sustained inside the target, 1.3 and 1.0 kg of tritium can be produced in a BWR and PWR respectively within one year. Stainless steel claddings, however, are known to be poor at containing tritium produced in fuel (see, e.g., Phillips and Easterly, 1980). Therefore it is necessary to find a target wall material which can sustain the same pressure as stainless steel but would release less tritium.

6. Magnetic confinement fusion neutron source.

   It has been shown that a helium-3 blanket would be an option for tritium breeding in near-term fusion devices. Some 25 to 50 kg of helium-3 would be needed as initial inventory. Based on 10 to 25% availability per year, 3.3–8.3 g/y of tritium could be produced (Steiner, 1989).

2.4.3 Boron path

The boron isotope ^{11}B undergoes a tritium-forming reaction on capture of one neutron[28]

$$^{11}B + n \rightarrow n + 2\alpha + {}^{3}H + 0.23\ \text{MeV}. \tag{2.9}$$

but this isotope does not contribute to the tritium production in a nuclear reactor since the threshold energy for this reaction is 9.55 MeV which is too high.

Some 19% of natural boron is ^{10}B, which produces tritium by the reaction[29]

$$^{10}B + n \rightarrow 2\alpha + {}^{3}H + 0.23\ \text{MeV}. \tag{2.10}$$

There will be an increase of the tritium production rate with target age due to buildup of ^{7}Li in the target by the nuclear reaction[30]

$$^{10}B + n \rightarrow \alpha + {}^{7}Li + 2.79\ \text{MeV} \tag{2.11}$$

which is followed by the tritium-producing reaction as given in Section 2.4.1 (lithium-path) (see Equation 2.3).

Diversion paths based on boron

1. Tritium production in movable boron carbide control rods in BWRs.
   The production in boron carbide control rods of BWRs has been estimated to be 0.2 $g/(GW_e y)$ in relatively new rods but increasing to 2 $g/(GW_e y)$ in well-burned rods. The average production is about 1 $g/(GW_e y)$ over a 15-year life (Smith and Gilbert, 1973). Probably, AGRs have the same production rate, whereas HWRs, MAGNOX, and HTRs have half this rate.[31] PWRs and FBRs do not produce tritium in the control rods.[32] Only about 0.2% of this appears to be released to the coolant (Peterson and Baker, 1985), i.e., most of it is retained in the rods and available for extraction.

2. Tritium production from boron as fixed burnable poison rods (BPR) in PWR.
   Boron carbide (BC_4) is normally chosen as the poison material and dispersed in 580 g of alumina (Al_2O_3) per rod at the dilution necessary to achieve the proper poison value. The total boron content per rod is about 7.5 g and about 10.8 kg in the whole core.[33] The tritium production rate is around 0.08 $g/(GW_e y)$ (Locante and Malinowski, 1973). Releases to the coolant are less than 1% (Peterson and Baker, 1985).
   After a residence time equivalent to 440 full-power days during a total of 707 days residence time and 1063 days cooling, 46.8 MBq tritium was found per gram pellet of a BPR. This value was confirmed by calculations (D'Annucci, 1982). Accordingly, for a 1 GW_e PWR[34] the total quantity of tritium at the time of reload from the reactor corresponds to a production rate of 1.1 $g/(GW_e y)$. 97.1% of the original boron content was found to be consumed.
   When BC_4 was used as poison material, 99.5% was found to be retained in the pellets, the rest in the zirkaloy cladding, and no tritium was detectable in the plenum gas (D'Annucci, 1982). However, between 30% and 80% of the tritium generated from B_2O_3 is estimated to be released to the coolant (Locante and Malinowski, 1973).

3. Tritium production in boron curtains, which are used to control excess reactivity of initial cores of BWRs, is around 0.2 $g/(GW_e y)$ (Kouts and Long, 1973). In newer BWRs these curtains are replaced with fuel tubes spiked with reactivity control material. 10–50% of the tritium generated in borosilicated glass absorber plates appears to be released to the coolant (Peterson and Baker, 1985).

2.4.4 Tritiated water path

The critical step to this path is not the production, since this in general occurs inadvertently. The critical step for tritiated water is the extraction of tritium. The main sources of tritiated water are the coolants and moderator of heavy water reactors and aqueous waste streams of reprocessing plants or tritium-handling facilities.

Diversion paths based on tritiated (heavy) water

1. Tritiated moderator or coolant of heavy water reactors.
   Tritium is produced inadvertently[35] in the moderator and coolant of a heavy

Table 2.3 Annual increase of tritium in heavy water reactors.

reactor type, part	specific production rate		reference
	[10^3 TBq/(GW$_e$y)]	[g/(GW$_e$y)]	
CANDU, coolant and moderator	88.8	247	(Gorman and Wong, 1979)
CANDU, coolant and moderator	70	195	(Stern, 1988)
CANDU, coolant and moderator	60–82	170–230	(CFFTP, 1988)
HWR, coolant	22.2	62	(Kouts and Long, 1973)
HWR, coolant and moderator	50–90	140–250	(Ebeling, 1985)
HWR, coolant and moderator	85	240	(Bonka, 1980)
	specific production rate		
	[GBq/(GW$_{th}$y)]	[g/(GW$_{th}$y)]	
FR2 Karlsruhe (44 MW$_{th}$)	22	60	(König, 1980)
MZFR Karlsruhe (200 MW$_{th}$)	10	30	(König, 1980)

water reactor during normal reactor operation through capture of a neutron by deuterium. The nuclear reaction is noted[36]

$$^2\mathrm{H} + n \rightarrow {}^3\mathrm{H} + \gamma. \tag{2.12}$$

Depending on the reactor type, the production rate via the heavy water path is (20-90)$\times 10^3$ TBq/(GW$_e$y), i.e., (60-250) g/(GW$_e$y) at 100% capacity (see Table 2.3).

In Tables A.2 and A.4 the annual increase of tritium is given for each country running heavy water reactors assuming a capacity factor of 100%. The accumulated amount can be calculated only if the history of reactor operation is known.

For example, for Pakistan's 125 MW$_e$ KANUPP reactor the annual increase is calculated as (20 to 30) g.[37] From the start of its operation in 1971 until September 1989, this reactor produced no more than 5126 GW$_e$h (Müller and Hossner, 1990) and a total of at most 150 g of tritium (not decay-corrected).

A moderator of a typical CANDU has 290 tonnes of heavy water. At equilibrium it would contain about 2.4 TBq/kg (i.e., the maximum total content is 700 GBq, i.e., 2 kg). The cooling system has a smaller quantity of heavy water (150 t), and it remains only part of the time in the neutron flux inside the reactor. At equilibrium the coolant would contain 0.09 TBq/kg (total content: 40 g). For radiation protection it is desired to keep the tritium content in the moderator low (e.g., below 0.2 TBq/kg).

(a) Divert tritiated heavy water for clandestine tritium extraction, and claim that the reduced tritium inventory is due to accidental losses into the environment. This implies that a plausible release scenario has to be invented.

If the concentration were 0.1 TBq/kg (maximum for coolant) a loss of 3.5 t would contain 1 g of tritium; if it were 2.4 TBq/kg (maximum for moderator) a loss of 150 kg would contain 1 g of tritium. In principle, coolant or moderator losses to this extent are realistic. For example, on April 18, 1989, 30 to 35 t of heavy water were lost due to a defective sealing ring at KANUPP in Pakistan. However, it is quite improbable that such losses are not kept within the reactor containment and available for inspection.

(b) Divert tritiated heavy water for clandestine tritium extraction and claim that the reduced tritium inventory is due to normal operational losses. This implies that emission monitoring data have to be falsified.

Leakage from reactor coolant is generally higher than loss of moderator because it is operated at higher pressures and temperatures. The total loss can be kept as low as 0.5% of the heavy water inventory, but up to 2 to 3% can be anticipated for large reactors. Assuming an optimum loss of 0.5%, the tritium release from a 1 GW_e HWR ranges from 100 TBq in the first year of operation to some 740 TBq ($\simeq$2 g) in the tenth. 80% is emitted as airborne effluents and 20% as liquid (UNSCEAR, 1977). For example, the average total release from the Pickering HWR site (1.63 GW_e) in Canada from 1971 to 1977 was 890 TBq/y (2.5 g/y) normalized to 1 GW_e (Gorman and Wong, 1979). The variation is larger over time because heavy water leakages have varying reasons.

Assuming a ten-year-old reactor which can be maintained at a loss of only 0.5%, a maximum diversion of another 2.5% could be claimed to be normal operational loss without raising suspicion. This would result in diversion of 10 g/y for a 1 GW_e HWR.

(c) Divert tritium without changing the inventory of heavy water.

Sufficiently slow replacement of tritiated heavy water in the heavy water reactor with pure heavy water in order to cheat the measuring device for tritium concentration. Since various estimates for the annual generation of tritium in the heavy water of a CANDU-type reactor vary by 80 g/(GWey) (see Table 2.3), this is about the maximum amount of tritium that can be diverted without detection.

2. Divert tritiated moderator or coolant water from reactors not using heavy water.

In the coolant of pressurized water reactors, tritium is produced from boron dissolved as burnable poison (as boric acid), activation of lithium present as an impurity or deliberately added for pH control (e.g., LiOH),[38] and from activation of deuterium.[39] These paths can yield some (0.072 to

Table 2.4 Production via fuel rod path.

reactor type	production rate [g/(GW$_{th}$y)]	reference	typical burnup in near future [GW$_{th}$d/t]	tritium in fuel, not decay-corrected [g/t]
PWR	1.6–2.1	(IAEA, 1981)	35–40	0.067–0.10
BWR	1.6–2.1	(IAEA, 1981)	30–35	0.057–0.87
FBR	2.1–3.1	(IAEA, 1981)	40–57	0.10–0.21
HWR	2.1–2.5	(IAEA, 1981)	9	0.022–0.027
MAGNOX	2.6	(Bonka, 1980)	18	0.056
AGR	1.8	(IAEA, 1981)	18	0.039
HTR	1.9	(IAEA, 1981)	100	0.23
HTR	1.2–1.8	(Phillips and Easterly, 1980)	100	0.14–0.21

0.089) g/(GW$_e$y).[40] More tritium may be released from burnable poison (30 to 80% in case of B_2O_3 but nearly nothing in case of BC_4) or fuel rods (no more than 1% in case of zircaloy claddings) to the coolant (Locante and Malinowski, 1973; and D'Annucci, 1982).

For BWRs this path plays no significant role. Only about 1 mg/(GW$_e$y) is produced in the coolant (Smith and Gilbert, 1973).

3. Extract tritium from tritiated waste streams that are diverted from reprocessing facilities.
   This is covered in detail in the next section.

4. Illegal extraction or removal of tritium at the detritiation plant.

   (a) Unreported detritiation of tritiated water from uncontrolled sources at an inspected (heavy) water detritiation facility. The largest facility in the world has a maximum extraction capacity of some 3 kg/y.

   (b) Illegal removal of tritium from the storage of extracted tritium at the (heavy) water detritiating facility combined with manipulating the inventory measurement so as to pretend to have a larger physical inventory which matches with the book inventory.

2.4.5 Ternary fission path

In the fuel rods of a nuclear reactor, tritium is produced inadvertently during normal operation mainly by ternary fission. In the course of this process, the nucleus fissions into two heavy and one light fraction. In one of about 10,000 fissions, tritium is produced. In addition, there is some tritium production from impurities in the fuel, especially from lithium-6. Depending on the reactor type, the production rate of tritium in fuel is 1.2 to 3.1 g/(GW$_e$y) (see Table 2.4).

Nearly all tritium produced in the fuel remains within the zirkaloy cladding (about 99%[41]) and will only be released by reprocessing or special heat treatment. Within the fuel rod the tritium is distributed in the cladding (5 to 15%), in the gas plenum (0 to 10%), and the rest in the fuel matrices (Phillips and Easterly, 1980). The fraction of tritium that finds its way into the aqueous phase depends on the design of the reprocessing plant. In the PUREX process, spent fuel is cut into pieces and dissolved in acid (see, for example, Neel, 1986). Half of the tritium from ternary fission and from lithium-6 impurities in fuel elements remains in the zircaloy cladding; the other half appears as tritiated water (HTO) or tritiated nitric acid (TNO_3) in the fuel solution (Bruggeman, 1985). Most of it (about 97%) is set free as gaseous tritiated water (HTO) emission in the following process in which the volume of nonvolatile radioactive isotopes is reduced by evaporization. Between 0.4% (oxide and carbide fuel) and 25% (metallic fuel) of all tritium is emitted in gaseous form as molecular hydrogen HT or T_2 while cutting and dissolving the fuel (Grathwohl, 1973).

Alternatively, the spent fuel can be heated before or after cutting to more than 1000°C to outgas most of the tritium (Bray, 1981; and Campbell and Pattison, 1981). A reduction of the tritium content in process streams by a factor of 100 could be achieved.

Nowadays, tritium from fuel rods is normally released into the environment, although complete extraction and storage is technically feasible. At present, only one small research facility is in operation in Belgium, which extracts tritium from tritiated reprocessing water (see Table A.11 and Bruggeman, 1985).

Without decay correction the tritium content of each ton of spent fuel from a LWR is about 0.06 to 0.10 g of tritium (see Table 2.4). Allowing for decay after a cooling time of 150 to 160 days and for some tritium remaining in wastes, the fraction of tritium available for release or recovering can be estimated to be 80% of the total tritium produced, i.e., 0.05 to 0.08 g/t of uranium.[42] From this it follows that in a reprocessing plant for LWR fuel with a yearly capacity of 12 t, a maximum of 1 g of tritium is recoverable. Commercial reprocessing facilities have typically a capacity which is larger by a factor of 50.

Diversion paths based on ternary fission

1. At a reprocessing facility, tritium-bearing off-gases are collected to extract tritium. Some 5 to 6% of the tritium which is contained in spent fuel is emitted in gaseous form from a PUREX LWR fuel reprocessing plant. If no tritiated waste water is evaporated to make use of the large mixing capacity of the atmosphere, the gaseous emission rate is about 0.005 g/t of uranium. A plant with a capacity of 200 t of uranium emits about 1 g/y.

2. Tritium can be clandestinely recovered from highly radioactive aqueous waste. In the case of PUREX it contains about 5% (Schwarz, 1975) and a maximum of 1 g for a 240 t/y facility.

3. Tritium can be clandestinely recovered from organic waste. In the case of PUREX it contains about 7% (Schwarz, 1975) and a maximum of 1 g for a 170 t/y facility.

4. Tritium can be clandestinely recovered from fuel cladding waste. In the case of PUREX it contains about 5% (Schwarz, 1975) and a maximum of 1 g for a 240 t/y facility.

5. At a reprocessing facility, tritium bearing aqueous effluents are collected to extract tritium. The fraction of the tritium emitted as aqueous effluent is typically some 80%, i.e., 1 g for a facility with a capacity of 15 t/y.

6. Fuel rods are diverted to extract tritium clandestinely.

7. Spent fuel rods are not diverted but are heated to extract the tritium.[43]

   It is technically feasible to extract tritium quantitatively from spent fuel without reprocessing. The fuel rods have to be opened and the fuel elements are heated in an oven either in a vacuum or with an inert sweeping gas.[44]

   A typical spent fuel assembly of a PWR has, for example, 236 fuel rods, each of which contains 2.3 kg of uranium. From the figures given above it follows that one SQ of tritium can be collected from some 25 PWR fuel assemblies weighing a total of 13,570 kg.

2.5 Removal path

2.5.1 Removal from tritium-handling facilities

When tritium is available in any chemical and physical form, there are numerous ways to remove it illegally for weapons purposes. In this analysis these paths are summarized in two different categories. One is the removal from storage or handling processes of pure tritium (this subsection); the other is the recovery of abandoned tritium (next subsection). For most of these paths the maximum divertable quantity depends on the total amount of tritium available and the relative accuracy of accountancy and inventory verification (see Section 3.4).

Removal of tritium from tritiated heavy water and from spent fuel are important paths of this category. They are dealt with in Sections 2.4.4 and 2.4.5, respectively. Other diversion paths based on illegal removal from stored or handled tritium at facility types 7 and 8 are given in the following list (for facility classification see Table 3.4):

1. Nonreported transfer of tritium from a "material balance area" (MBA)

2. Partial discharge of storages during transport

3. Overstated tritium content of waste

4. Claimed release to the environment

5. Only part (e.g., 80%) of declared quantity of tritium put into radio-luminous paint or other products

6. Divert accumulated "hidden inventory"

2.5.2 Recovery of abandoned tritium and multisource acquisition

Recovery of tritium from tritiated heavy water and from spent fuel are the major paths of this category. They are described in Sections 2.4.4 and 2.4.5, respectively. Other paths of tritium recovery and accumulation from multiple sources are given in the following list:

1. Clandestine recovery and purification of tritium from solid or liquid waste after termination of safeguards.

   For example, about 15 m^3 cladding waste is produced per $GW_e y$. About 5% of the tritium generated in the fuel remains in the cladding, i.e., about 0.1 $g/(GW_e y)$ for LWRs. Therefore, the concentration is about 6.7 mg/m^3.

2. Clandestine recovery and purification of tritium from aqueous emissions.[45]

   In facilities handling tritium, typical annual losses to the environment are in the order of 1% of the inventory or throughput. For example, the tritium release fraction from the separations area at the Savannah River Plant was relatively constant at 0.9% over the lifetime of the facility (at least from 1955 to 1981). The annual throughput ranged from 2 to 10 kg, i.e., between 20 and 100 g was lost to the environment each year (Cochran, 1987a).

3. Clandestine recovery and purification of tritium from gaseous emissions.[46]

   The same argument as in the previous point applies. The amount of gaseous emissions depends on design specifications. Tritium emissions from nuclear waste disposal are calculated in gaseous form. The application value for tritium emission from Schacht Konrad is 30 TBq/a.[47] This is less than 0.1g/y.

   It is questionable how much tritium can be gained from gases. Tritium removal systems developed for radiation protection purposes are able to clean air with concentrations of more than 1 MBq/m^3.

4. Clandestine recovery of tritium generated in magnetic confinement fusion experiments.

   Tokamak experiments are normally made with deuterium gas. The total production of tritium by deuterium–deuterium fusion is estimated to be 0.0001 g/y.[48]

5. Clandestine recovery and purification of tritium from consumer products after termination of controls.

   The estimate of the tritium quantity that could be gained from recycling is similar to that of the preceding point. However, it is difficult to imagine how this could be done clandestinely with a large number of such consumer items which contain small quantities of tritium. For example, the 62 g of tritium sold by the DOE in 1986 to domestic civilian customers were supplied to 16 different companies and research institutes in 130 shipments (Stern, 1988). In the same year, the Nuclear Regulatory Commission (NRC) licensed nine tritium exports from the U.S. containing more than one Ci. The total of exports was 1.37×10^6 Ci (137 g).

Nevertheless, making accountable all products which contain more than a certain limit of tritium each should be considered. For example, remote airfield runway lights contain up to 0.1 g, but most other products contain less than 10 mg (see Table 1.3).

The useful life of a tritium gas-filled light source is from 8 to 10 years. After this time, about 40 to 50% of the original charge will have decayed due to the tritium half-life of 12.3 years. The remainder is available for recovery and reuse. The recyclability will certainly depend on the form in which tritium is introduced in the product. It can be dispersed in paint or plastic, can be filled in sealed glass tubes, or can be contained in other ways.

The radio-luminescent (RL) lights program at the Oak Ridge National Laboratory (ORNL) is developing a tritium recovery system with an ultimate goal of transferring the technology to private industry in the RL light manufacturing business. Over 2.2 PBq (6.2 g) of tritium exist in an unusable form in rejected light sources at ORNL (Kobisk, 1989).

6. Receiving many shipments with each less than the accountable amount.

2.6 Survey of worldwide civilian stocks and production capacities

The facility types referred to in this section follow the same scheme used in Table 3.4 and in Appendix A. Most of the data presented here are as of 1992. No significant changes have occurred since then. All considerations regarding tritium control that are based on these historic numbers are still fully valid.

2.6.1 Nuclear reactors and special neutron sources (facility type 1)

Nuclear power reactors (facility type 1a,b)

Table A.1 lists all nuclear power reactors (including heavy water reactors) ordered by country. The quantity of tritium produced in heavy water reactors due to neutron capture of deuterium (heavy water path) is given in Table A.2. Table A.1 indicates the quantity of tritium which is produced inadvertently by ternary fission and which could be produced by breeding from lithium-6 targets without affecting normal operation. It should be noted that this specific production rate is much smaller than it would be if the reactor were converted to a dedicated production reactor.

The amount of tritium produced inadvertently by ternary fission in nuclear power reactors is estimated here to be some 150 PBq/y (410 g/y) assuming a worldwide reactor capacity of 330 $GW_e y$ at the end of 1992, working at 100% capacity factor. Another study estimated the tritium production by power reactors to be 105 PBq/y (290 g/y) during 1984 while generating about 134 $GW_e y$ (Luykx and Fraser, 1986).

The expected increase in nuclear power generation by the year 2010 is at most 10%. Thus, worldwide annual quantities of tritium produced in power reactors will increase at most by a similar low rate.

Table A.2 lists all heavy water power reactors ordered by country. It indicates the quantity of tritium which is produced inadvertently by capture of neutrons in deuterium, namely 1080 to 1600 PBq/y (3000 to 4400 g/y). Other production paths in these reactors are covered in Table A.1.

Nuclear research reactors (facility type 1c,d)

Table A.3 lists all research reactors ordered by country. It gives the quantity of tritium which is produced inadvertently by ternary fission, i.e., only 0.92 to 1.8 g/y. It also shows the amount of tritium which could be produced by dedicating the reactors to tritium production and breeding it from lithium-6 targets.

Table A.4 lists all heavy water research reactors ordered by country. It indicates the quantity of tritium which is produced inadvertently by capture of neutrons in deuterium, namely 5.6–23 g/y.

Military production reactors (facility type 1e)

This facility type is covered in Section 2.7.

Special neutron sources (facility type 1f)

Nuclear reactors are the only continuous neutron sources which are of relevance for tritium production. They are all covered in the two previous paragraphs. Several techniques are available to generate pulsed neutron beams (see list in Appendix A.1.3). The only technology able to produce average neutron fluxes high enough to enable breeding of significant quantities of tritium are spallation neutron sources (SNS; see Table A.5). For example, the average neutron flux of SNS is lower than the peak flux by a factor of 650. In the U.S., SNS is considered as a possible future tritium production facility.

Magnetic confinement fusion devices are intended to breed large amounts of tritium. In order to breed as much tritium as it burns, a 1 GW_e fusion power reactor (if existing and working at full capacity) would need to produce about 170 kg tritium per year.

2.6.2 Fuel fabrication facilities (facility type 2)

Facilities of this type do not contain any tritium. They are relevant because the raw material for tritium breeding, basically lithium-6, might be inserted in fuel assemblies at these facilities.

In 1993, 44 fuel fabrication plants were in operation in 19 countries. The total annual capacity was more than 16,000 tonnes of heavy metals.[49] The capacity for light water reactors was larger than 11,000 t/a. A number of additional countries are going to install domestic fuel fabrication capacities (Egypt, Turkey, and Yugoslavia) (IAEA, 1989). Forty-three fuel fabrication facilities are under IAEA safeguards in 18 countries (see Table A.6). This includes some of the above-mentioned 44 plants more than once because some plants are composed of more than one facility. In addition, this figure includes some lab-scale facilities. At least three plants are not under safeguards in nuclear threshold countries (India [BARC], Pakistan[50]). A list of fuel fabrication plants is given in Table A.7.

The core of a 1 GW_e light water reactor contains about 100 tonnes of fuel, a third of which is replaced each year. Fuel assemblies for PWRs consists of, for example, 20 control rods and 236 fuel rods, each containing 2.3 kg of uranium (i.e., 540 kg for the complete fuel assembly, 185 fuel assemblies for a complete core). Fuel assemblies for BWRs have, for example, fuel rods with 3 kg each (a total of 190 kg, 530 assemblies). Fuel assemblies for FBRs may have 166 fuel rods with 170 g each (a total of 28 kg).

A fuel fabrication plant of medium size (300 to 360 t/y) produces about 600 fuel rods per day, i.e., either 2 to 3 PWR fuel assemblies per day or 9 to 10 for a BWR.

2.6.3 Separate storages for spent fuel (facility type 3)

At present there are only interim storage facilities and no final disposal sites for spent fuel. The latter would be treated in Subsection 2.6.5 on nuclear waste disposal sites. In 1988, the world's total amount of spent fuel was about 60,000 tonnes of heavy metals. The annual increase was about 6000 t in 1988 and estimated to be 9400 t in 1995.[51] Figures given in Table A.8 are somewhat lower (>73,000 t in 1992), because reprocessed fuel as well as fuel stored in reactor pools is not included. The cumulative amount of unreprocessed spent fuel at the end of 1994 can be estimated to be >100,000 t. By 1992, some 40,000 t of spent fuel have been reprocessed and in 1991 about 5000 t (see Table A.9). The tritium content in spent fuel depends on fuel type and burnup (see Table 2.4). For all fuel reprocessed by 1992, the average tritium content due to ternary fission was 0.015 g/t (not corrected for decay). Due to higher burnups this figure will go up in the future to some 0.055 to 0.075 g/t. By end of 1994, a total of more than 4 kg of tritium had been generated in fuel by ternary fission. A fraction of this is released by reprocessing (0.6 kg) (see Table A.9) or lost by decay. The remainder is stored in unreprocessed spent fuel. The yearly increase of tritium in spent fuel is around 0.41 kg (see Table A.1).

Most spent fuel is stored at the reactor site. Besides at-reactor pool storages, a number of countries built away from reactor storage (AFRS) capabilities. The current AFRS capacity is about 27,000 t and an additional capacity is under construction (IAEA, 1989). In Germany, the Brennelementlager Gorleben (BLG) (in operation since 1984) has a capacity of 1500 t, and the Brennelement-Zwischenlager Ahaus (BZA) has been in operation since 1991 with the same capacity (Müller and Hossner, 1991). Each contains up to some 200 g of tritium from ternary fission (not corrected for decay).

For separate spent fuel storage facilities (at the reactor or away from the reactor) under IAEA safeguards, see Table A.8. The total capacity of 38 facilities (partly planned) is larger than 100,000 t. Of these facilities, 23 are currently under international nuclear safeguards.

2.6.4 Reprocessing plants (facility type 4)

By 1991, some 40,000 t had been reprocessed worldwide, releasing some 0.48 kg of tritium to the environment.[52] In 2005, some 8,000 t of spent fuel may be reprocessed annually, releasing in one year between 0.36 and 0.48 kg of tritium to the environment.[53]

Normally, two parallel reprocessing lines of identical capacity are built, but only one is operated at a time. Work is normally performed in campaigns and not in continuous operation. Therefore, the design capacities do not adequately reflect the quantities actually reprocessed. All the commercial reprocessing plants of the world are listed in Table A.9.

The specific content of tritium in fuel of some 0.05 to 0.08 g/t implies that reprocessing facilities with an annual capacity of a little more than 12 t can have an annual throughput of more than 1 g of tritium. Currently, there are 10 of these facilities in operation worldwide.

In addition, there have been major reprocessing activities associated with weapons production. However, the burnup is significantly smaller and therefore less tritium was produced during plutonium breeding than by power reactor operation. Some 300,000 to 800,000 t of uranium have been reprocessed with low burnup (about 600 MWd/tU) to extract plutonium in military facilities worldwide. Accordingly, between 0.45 and 1.2 kg of tritium have been released.[54]

2.6.5 Final disposal sites for nuclear waste (facility type 5)

In Table A.10 all existing and projected final (i.e., no interim) nuclear waste disposal sites are listed. In 17 countries, more than 26 facilities are in operation or already closed, and at least 42 facilities are planned in many countries, including 16 which do not presently have an operational facility. Some countries (Canada, Sweden, U.S.) plan to keep spent nuclear fuel for a long period of time (50 to 100 years) in interim storage facilities. At present, no nuclear waste disposal facilities are under safeguards except for some inspection activities at the Mol site in Belgium.

Relevant tritium quantities can be expected only in final disposal sites for high-level radioactive wastes. According to present plans, no appropriate disposal sites will be in operation before 2010.

Waste in final repositories is by definition irretrievable.[55] Hence, it is not of severe concern in terms of nonproliferation that the total content of tritium at the end of the operating phase can be large. For Schacht Konrad (Germany) it is expected to be 6 $\times 10^5$ TBq (1670 g) (Bundesamt für Strahlenschutz (BfS), 1990). Final disposal sites will play a major role in terminating safeguards. However, the half-lives of plutonium and uranium-235 are too large to be used as a reason for termination. Irretrievability is required as a criterion. This is different for tritium.

2.6.6 Detritiation facilities (facility type 6)

So far, only six large-scale industrial or pilot research facilities which extract tritium from tritiated heavy water are in operation worldwide (see Table A.11).

The tritium removal plant at Darlington (Canada), which is operated by Ontario Hydro, is the world's largest nonmilitary producer of tritium. It extracts tritium from the heavy water which is used in Ontario Hydro's CANDU reactors. The total tritium recovery by the year 2001 (decay-corrected) was expected to be 23 kg, assuming a capacity factor of 80% (CFFTP, 1988). In fact, in 2003, an inventory of 18.5 kg separated tritium was achieved and the annual recovery rate had fallen from 2.1 kg/y to 1.5 kg/y (Willms, 2003).

The second-largest facility is located in Canada as well (capacity: 140 g/y); the next smaller facility is in Grenoble, France (16 g/y). The capacity of the Indian detritiation facility has not been published. A small research facility is in operation in Belgium, which extracts tritium from tritiated reprocessing water (Bruggeman, 1985). The smallest facility is the Tritium Aqueous Waste Recovery System (TAWRS) at Mound, OH, U.S. (0.06 g/y).

2.6.7 Tritium storages and research facilities (facility type 7)

For a list of tritium storage and research facilities, see Table A.12. There are 23 facilities in operation with a total inventory of >400 g, which may increase to 1 kg. In 17 facilities the inventory is larger than 1 g. Nine of these are in the U.S., five in Europe, two in Canada, and one in Japan. An additional seven facilities are planned. If fusion energy research makes significant progress, the maximum inventory may well rise beyond 5 kg.

A classification of tritium laboratories is difficult because several parameters have to be taken into account. With respect to radiological safety, a classification was suggested by the International Commission on Radiological Protection (ICRP) based on the activity inventory I[56] and another by the IAEA based on the tritium inventory.[57] In this study the relevant parameter is A = max(I,T), i.e., the maximum of inventory I and annual throughput T. The category which is significant with respect to tritium controls is defined in this monograph by A > 1 g. All tritium research facilities falling in this category are basically related to fusion energy research.

There is a definition problem, because there are single small portable containers (uranium getter storages, steel containers, or glass ampoules) which contain more than 1 g each. However, it seems reasonable not to include such items in this list. Only complete facilities with all the necessary infrastructure for safe tritium handling are considered. The reason for this is that such facilities have to pass a license procedure according to the relevant radiation protection legislation before they can receive an amount of tritium, which has to be below or equal to the licensed quantity.[58] The assumption is that no declared facility without such a license will be able to handle safely 1 g of tritium.

2.6.8 Tritium industry (facility type 8)

This facility type is defined by inventory and annual throughput just as described in the previous paragraph. For some large commercial tritium manufacturers and trade companies, see Table A.13. There are 21 large commercial tritium manufacturing and trading facilities worldwide, which have a total annual throughput of more than 300 g. That is the bulk of the whole world market in tritium. Only four of these companies are in nonnuclear weapons states (Canada, Germany, and two in Switzerland).

Table 2.5 Summary of annual tritium production, throughput rates, and stored inventories.

facility type by numbers as in Table 3.4 / main production or flow path	annual production or throughput rate [kg/y]			cumulative production or stored inventory decay-corrected [kg]		
	i: inadver. d: deliber.	c: collected r: released t: throughput	*potential production*	i: inadver. d: deliber.	c: collected r: released s: stored inv.	*still extractable*
1a,b / ternary fission	i 0.45	r<0.01	–	i 3.2–8.8	r 0.03–0.09	*2–7*
/ lithium	d 0	–	*10–70*	–	–	–
1b,d / heavy water	i 3.0–4.4	r<0.2	–	i $\simeq$20	r $\simeq$ 2	$\simeq$ *13*
1c,d / ternary fission	i 0.002–0.004	r<10^{-4}	–	–	–	–
/ lithium	d $\le$0.5	–	*10–20*	–	–	–
1e, U.S. / lithium	d 0	–	*2.5–3.0* [a]	d 70±25 [b]	c 68±25 [c] r 0.2–0.5	$\sim$*0.7*
/ heavy water	d 0	–	–	i 1–2	c 0.05–0.1	$\sim$*1–2*
1e, Russia / lithium	d 0–3.4	c 0–3.4	*4* [d]	d 66 [e]	c 65[f] r 0.2–0.5	$\sim$*0.5*
1e, others[g] / lithium	d<0.25	c<0.25	3 [h]	d 2.5–5	c 2.5–5 r 0.1–0.2	$\sim$*0.05*
1f, 1 GW_e fusion reactor, self-sustaining tritium breeding (fiction) / lithium	–	–	*180*	–	–	–
3 / storage	–	t 0–1	–	–	s 1–3	–
4 / split to release and waste[i]	–	t 0.20–0.33	–	–	r 1–1.4	
5 / storage	–	t 0.1–0.3	–	–	s 0.5–2	–
6 / extraction	–	c 2.0±0.5	–	–	s 6.2±0.2	–
7 / storage	–	t 0.03–0.05	–	–	s 0.4–0.6	–
8 / manufacturing	–	t 0.3–0.4	–	–	s 0.05–0.1	–
environment / nuclear explosions	–	–	–	i 300–3000	r 30–300	*0*
environment / natural production	i 0.1–0.4	–	–	i 1.5–6.7	–	*0*
total	2–8.6			450–3000		

[a] The last tritium-producing reactor at the Savannah River Plant was shut down in April 1988. The new tritium production at the Watts Bar pressurized water reactor will start in 2003 at a rate of 2.5 kg per year (DOE, 1999).

[b] Production rate and stockpile estimates of Cochran *et al.* (1987) are extrapolated to the end of 1993. By end of 2003 this inventory has decayed to 40 kg.

[c] Total amount in the nuclear stockpiles and production pipeline. At least 1% of the produced quantity is lost to the environment, and about 1% is expected to be in radioactive waste.

[d] See Table 2.6.

[e] Production rate and stockpile estimates of Cochran and Norris (1993) are extrapolated to the end of 1993 assuming annual production of 3.4 kg. There is not much sense in estimating the error of these figures because they are educated guesses.

[f] Some 3 to 5 kg of this inventory has been with the nuclear warheads which remained on the territory of the Ukraine at the end of 1993 and some 3 to 4 kg in Kazakhstan. There was some tritium in Belorussia, too.

[g] This includes the other nuclear weapons states, China, France, and U.K. as well as the defacto nuclear weapons states India, Israel, and Pakistan.

[h] See Table 2.6.

[i] See stored inventory in facility type 5 (waste storages).

2.6.9 Summary and outlook

In order to assess the proliferation risk posed by various stocks of tritium in different facilities, it is of interest to develop a picture of tritium flows and inventories at various places. Table 2.5 presents a survey of the total annual world tritium production rates (inadvertently, deliberately, potential) and the inventories in different facility types and in nature. Included are military facilities as facility type 1e, summarizing the data given in Table 2.6. The facility types and their worldwide number are given in Table 3.4.

The distribution of the main changes in civilian inventories (countries with more than 1 g/y) over the globe are depicted in Figure 2.5.

From this summary it is obvious that a nuclear reactor can be regarded as the bottleneck for today's tritium production. Some kilograms are still believed to be produced and extracted for military purposes each year by using the lithium path in dedicated production reactors. As to civilian activities, the largest quantities are released by reprocessing of spent fuel, and the largest quantities are collected by detritiation of heavy water. The heavy water path provides the largest civilian source of tritium. There exists worldwide a capacity to extract and collect yearly up to 2.5 kilograms of tritium from tritiated heavy water.

2.7 Survey of worldwide military tritium production

In general, information about military production of tritium is kept secret. Some data about dedicated military production facilities are available in the open literature and summarized in Table 2.6. This table includes all plutonium production reactors, since tritium production is in competition with plutonium production and can be coproduced with plutonium in the same reactor at any percentage from 0 to 100.

Recognized nuclear weapons states. For the U.S. the inventory at the end of 2003 can be estimated at (40±14) kg.[59] In April 1988, the K-Reactor, which was the last U.S. source for military tritium and which was exclusively dedicated to tritium production, was shut down, and no fresh tritium has since been produced for the U.S. nuclear arsenal. In August 1988, the U.S. decided that all remaining production reactors would have to undergo significant upgrading for safety reasons before they could be restarted. The U.S. spent $2.345 billion between FYs 1989 and 1992 trying to restart tritium production in the K-Reactor, without success.

In September 1992, then DOE Secretary James D. Watson said that the U.S. could recycle enough tritium from dismantled warheads to supply a reduced U.S. nuclear arsenal until 2012. According to an announcement by the succeeding DOE Secretary Hazel O'Leary in March 1993, the K-Reactor would not be restarted and placed on "cold standby." The program to design a new production reactor (NPR) was officially stopped by the Clinton Administration. However, in June 1993, O'Leary declared that a new production source for tritium should begin operation

Table 2.6 Military production reactors used or usable for plutonium and tritium production (facility type 1e).

country	facilities	years of operation or tritium production	production capacity for H-3 [g/y] [a]
China [b]	Second Ministry of Machine Building Industry	since 1968	?
	entirely new (larger) production line	since 1979	?
France	Marcoule, G1 (40 MW_{th}), G2 (250 MW_{th}), G3 (250 MW_{th}) [c]	since 1956, 59, 60, until 1968, 80, 84	only Pu
	Celestin I/II, 250 MW_{th} heavy water reactors [d]	since 1967, 68	2 × 750 [e]
	accelerator-based production facility	under consideration	?
India [f]	Cirus (BARC), 40 MW_{th} heavy water reactor	since 1960	120
	Dhruva (BARC), 100 MW_{th} heavy water reactor	since 1988	300
Israel	Negev Nuclear Research Center Dimona: IRR-2 (150 MW_{th}) [g]	since 1963	500
Russia [h]	Ozersk (former Chelyabinsk-65)	first since 1948,	little [i]
	5 graphite-moderated water-cooled reactors (total $\simeq$ 6565 MW_{th})	all shut down by end of 90	
	heavy water reactor, ~50 MW_{th}	~1950 until late 1980s	55 [k]
	2 light water reactors Ruslan and Lyudmila, ~1000 MW_{th} each	still operational	~4000 [l]
	Seversk (former Tomsk-7)	since early 1960s,	only Pu (?) [m]
	5 graphite-moderated water-cooled reactors, ~2000 MW_{th} each	3 shut down by end of 1992, last 2 to be shut down by 2008	
	Zheleznogorsk (former Krasnoyarsk-26),	since 1957, 61, 64,	only Pu [n]
	3 underground graphite-moderated reactors, ~2000 MW_{th} each	2 shut down in 1992, last one to be shut down by 2011	
U.K. [o]	Windscale, two 115 MW_{th} reactors	shut down	only Pu
	Calder Hall, four 220 MW_{th} Magnox reactors	since 1956, 58	only Pu
	Chapelcross, four 220 MW_{th} Magnox reactors	since 1958, 60	4 × 200 [p]

Table 2.6 *(continued)*

country	facilities	years of operation or tritium production	production capacity for H-3 [g/y] [a]
U.S. [q]	Hanford (Washington), nine reactors, including	1952–1988	mainly Pu
	N-Reactor, 4800 MW_{th}, graphite-moderated, light water cooled	short period ~1967	6250 [r]
	Savannah River Plant (South Carolina), five reactors, including	shut down	mainly Pu [s]
	K-Reactor, 2400 MW_{th} heavy water cooled and moderated	1953–1988	6300–11500
	Watts Bar, pressurized water reactor, ~1200 MW_{th}	start-up 2003	3000 [t]
	new accelerator-based production facility	under development	?
total	> 20 facilities operating, 28 shut down, 2 planned		

[a] Figures are very rough estimates of the maximum production rate without simultaneous plutonium production if running at full capacity and totally dedicated to tritium production, in some cases with simultaneous power production.

[b] JPRS (1988).

[c] See Cochran *et al.* (1987). Tritium breeding from lithium-6 started before 1962. After completion of the two Celestin reactors, G1, G2, and G3 were used for plutonium production only. See CEA (1962), p. 129.

[d] See Hugony *et al.* (1973) and Barrillot (1991).

[e] Alternatively, 45 kg plutonium could be produced according to Gsponer (1984). This estimate for tritium production appears to be far too high, probably by one order of magnitude. In fact, in 1980 these two reactors were modified to produce plutonium and less tritium (Barrillot, 1991). In 1992, the additional production of plutonium was terminated. Since that time, these reaction alternate in operation, presumably producing tritium

[f] See Albright and Zamora (1989). Both reactors are unsafeguarded and it is not clear whether they are used as production reactors.

[g] Upgraded from 24 to 150 MW_{th} in 1969. See Arms Control Reporter 13,5 (1993) 453.E.1

[h] See Cochran and Norris (1993), pp. 45, 47, and the Arms Control Reporter 12, 4 (1993) 611.E-0.4

[i] There was no major tritium production at these facilities. Sometimes tritium was produced in control rods (NRDC, 1989).

[k] Value is given in a CIA study which was conducted in the mid-50s and is quoted in Cochran *et al.* (1989).

[l] These two reactors are used for the production of tritium, ^{238}Pu and other isotopes.

[m] Tritium production is not known.

[n] Tritium has never been produced here. The remaining reactor is simultaneously used for power generation.

[o] Cochran *et al.* (1987). Shut down of all remaining eight reactors is considered for the period 2008–2010.

[p] Dual-purpose Magnox reactors producing electricity and plutonium or tritium.

[q] Cochran *et al.* (1987).

[r] With simultaneous production of power and 815 kg plutonium (Ragheb, 1981).

[s] Tritium was produced mainly in C-, later in K-Reactor. The restart for the latter was scheduled several times but never successful. In 1993, it was decided to keep the reactor permanently shut down.

[t] See DOE (1999).

in 2008. To meet this date, construction of an NPR would have to begin by the end of the 20th century. Alternatively, construction of a proton linear accelerator would have to start in 2002. Some scientists hoped to have developed the technology to produce tritium with an accelerator to maturity in due time.

Eventually, at the end of 1998, the new DOE Secretary, Bill Richardson, decided[60] to contract the commercial nuclear power plant operator, Tennessee Valley Authority (TVA), to use its pressurized water reactors, Watts Bar and Sequoyah (1200 MW_{th}), to produce tritium for military purposes. Of historic significance is the decision to abandon the policy of keeping the military and "peaceful" uses of nuclear energy separated (Bergeron, 2002). The production capacity will be on average 2.5 kg and up to 3 kg of tritium (surge capacity) per year. The first charge of burnable absorber rods containing lithium-6 are scheduled for placement into the reactor core in September 2003. After an irradiation period of 18 months, tritium will be extracted from the breeding targets at the new tritium extraction facility under construction at the Savannah River site. The DOE continues its dual-track approach. The design of the accelerator-based production of tritium will be completed as a backup technology. The construction of this facility will not be initiated.

At that time, the operational U.S. stockpile consisted of about 9250 strategic and tactical warheads and bombs and was planned to be reduced according to START II to about 4450 by 2003 (Norris and Arkin, 1994). The tritium requirement for the post-START II arsenal would have been about 9 to 13 kg. After implementation of the Moscow Treaty by 2012, the tritium demand will be about 5 to 10 kg, allowing for 1700 to 2200 strategic and 1000 tactical nuclear weapons. In 2003, the remaining tritium stockpile can be estimated at 26 to 54 kg. Depending on which combination of figures describes reality correctly, the year in which the U.S. tritium stocks decay below the demand may be as early as 2020 or as late as 2045.[61] If a stockpile of 1000 warheads were considered sufficient, existing tritium stocks could last until 2040 or even 2060.

Though the military tritium inventory has never been declassified, it can be estimated from information about the planned production rate as published in DOE (1999). Accordingly, approximately 2.5 kg per year will have to be produced, starting in 2003, in order to replace tritium that will have decayed in the total required inventory. Therefore, the inventory available to support requirements for the nuclear weapons stockpile on pre-START II levels and a five-year tritium reserve will be at least 45 kg in 2005 when the first batch of freshly produced tritium becomes available.

The tritium-producing reactors of other countries are likely to face a similar fate to that of U.S. reactors. Most of them have been in operation for 25 to 35 years (see Table 2.6). Russia has already shut down 11 of its 14 plutonium production reactors and pledged in June 1994 to phase out the operation of the remaining three reactors by the year 2000 (Gore–Chernomyrdin Agreement of 1997). In March 2003, the deadline was extended to 2008 for two reactors and 2011 for the last one. One source reports that two additional light water reactors are dedicated to the production of tritium and other isotopes (Cochran and Norris, 1993, pp. 45–47). In 1989, Soviet officials said that their country would have a continuing requirement for two to three tritium production reactors (Cochran and Norris, 1993, p. 43). Although no official declaration regarding tritium production has been made, this estimate has certainly changed because of the continuation of the nuclear disarmament process in Russia, especially the unilateral declarations by Yeltzin in 1992 and the ratification of Moscow Treaty in 2003.

The current annual production of tritium in Russia can be assumed to be between zero and a few kg. One educated estimate arrived at a decay-corrected cumulative stock of 66.3 kg at the end of 1991.[62] The decay-corrected worldwide military inventory of tritium can be estimated at about (140±30) kg at the end of 1993 (see Table 2.5). Without of any further production this inventory would have decayed down to 80±17 kg by the end of 2003.

States with nuclear weapons programs not recognized by the NPT. There are indications that most countries which are known or suspected of having developed nuclear weapon capabilities have also engaged in acquiring tritium and tritium technology to enhance these capabilities. The following states are suspected or known of using or having used tritium within their declared or suspected weapons programs: India, Iraq, Israel, Pakistan, and South Africa. Their tritium-related activities are discussed in Section 1.7. India and Israel are included in Table 2.6.

2.8 Conclusions on tritium diversion

2.8.1 Assessment of diversion possibilities

The following paths to acquire significant quantities of tritium within one year are possible: breeding of tritium (lithium-6, helium-3, and boron-path), extraction of inadvertently produced tritium (heavy water path and fuel rod path), and removal of stored pure or recoverable tritium (tritiated heavy water, spent fuel, tritiated wastes, consumer products). For each of these paths there exists a variety of technical alternatives depending on the facilities concerned. A summary of all significant paths and their typical or maximum diversion rates is given in Table 3.5.

Significant amounts of tritium can be produced in any high-flux neutron source, especially in any nuclear reactor. As these neutron sources are large plants which cannot be hidden easily, it seems to be feasible to implement a tritium control system which puts its main effort on a few hundred neutron sources in the world.

The rate at which a target material is transformed to become tritium depends upon the available neutron flux and the reaction cross-section of the breeding reaction. Figure 2.4 shows the percentage of target materials that is transformed as a function of exposure time in a neutron flux of 10^{13} $s^{-1}\,cm^{-2}$. These breeding rates are optimistic because this flux is very high and not reached in most or all parts of the power reactor core. Furthermore, an infinitely thin target has been assumed for the calculation, i.e., the flux depression inside the target is not taken into account.

Though helium-3 has the largest cross-section of the four materials compared in Figure 2.4 and 60% of the target material is transformed to tritium within one year, it has not yet been of practical significance for tritium production because it is a gas and it is technically difficult to expose significant quantities to the neutron flux.

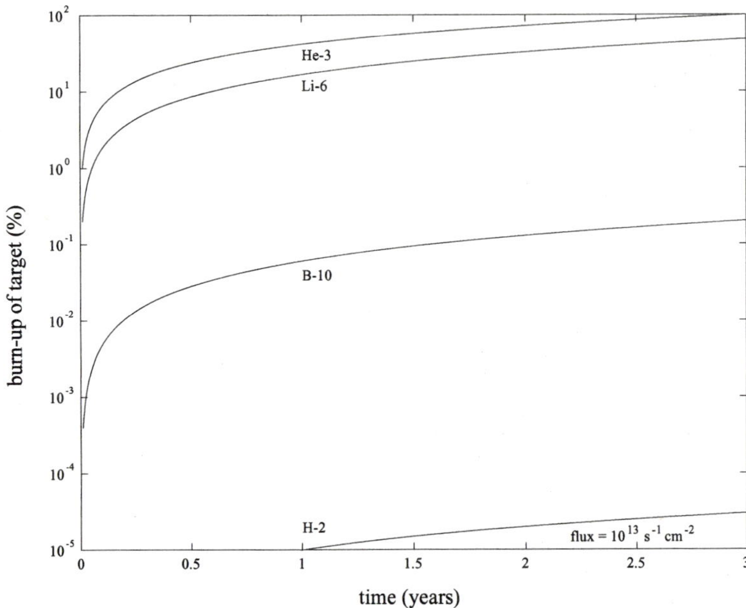

Figure 2.4 Tritium breeding from lithium-6, helium-3, deuterium, and boron-10.

Lithium-6 is the most important raw material for tritium breeding, and about 20% of the target material can be burned within one year at a neutron flux of 10^{13} $s^{-1}cm^{-2}$.

In spite of its low cross-section, deuterium is the second-most relevant raw material for tritium production because large quantities of heavy water are exposed to the neutron flux in heavy water reactors.

In Table 2.7, the specific production rates for the relevant deliberate and inadvertent production paths in a fission reactor are given. **From this table it becomes clear that for anybody who wants to produce large amounts of tritium the method of choice is the lithium-6 path.** This method is followed by all the nuclear weapons states.

Helium-3 targets are considered for future accelerator-based breeding systems. But this option, although technically feasible, is not currently being followed on a significant scale.

If by chance (or just for this purpose) heavy water reactors are in place, detritiation of the heavy water will constitute the best alternative. Canada is the largest producer of tritium for civilian purposes. It extracted yearly up to 2.5 kilograms of tritium from the heavy water moderator and coolant of up to 21 CANDU power reactors. In 2003, the production rate was down to 1.5 kg/y.

From this diversion path analysis it is obvious that a nuclear reactor can be regarded as the bottleneck for today's tritium production. Control measures could likewise be confined to nuclear reactors and in addition to some related facility types such as fuel fabrication plants as well as any facilities containing significant quantities or throughputs of tritium.

Table 2.7 Specific production rates for various schemes to produce tritium in a fission reactor.

path	production mode	[g/(GW$_{th}$y)]
lithium-6 path	dedicated production reactor	1000–5000
	LMFBR, lithium coolant instead of sodium	1000 [a]
	LWR or CANDU reactor with Li-targets, without affecting normal operation	30–70 [b]
	PWR, with 600 to 900 target rods in free control rod guide tubes	30 [c]
	PWR, burnable poison rods containing ^{6}Li instead of B or Gd	2–27 [d]
	LWR reactor with single Li-target rod in fuel assembly	1
	power reactor, inadvertent production due to ~.05 ppm ^{6}Li in fuel	0.003–0.3
fuel rod path	inadvertent production by ternary fission	0.5–1.0
heavy water path	inadvertent production by neutron capture in heavy water moderator and coolant	50–80
helium-3 path	experimental reactor loop, NRX reactor at CRNL	4.3 [e]
	research reactor, rapid power excursion experiments	<5 [f]
	HTGR, inadvertently production in helium coolant	0.06–0.23 [g]

[a] Ragheb (1981).
[b] CFFTP (1988).
[c] Derived from data given in Lu *et al.* (1988).
[d] Benedict *et al.* (1981) for the lower and Ragheb (1981) for the upper value.
[e] Derived from data given in Osborne (1979).
[f] Sokolski (1982).
[g] Phillips and Easterly (1980).

The easiest way to divert tritium is the illegal removal from existing stocks of tritium. However, there are not many opportunities to collect significant quantities by illegal removal of pure tritium because there are very few facilities worldwide which handle more than one gram of tritium within one year. Larger quantities of tritium are available in heavy water or in spent fuel. But the number of facilities in the world in which these resources of tritium are extractable is very small. Thus, it appears to be a fairly limited task to verify the nonremoval of tritium.

In order to assess the proliferation risk posed by various stocks of tritium in different facilities, it is of interest to develop a picture of tritium flows and inventories at various places. Table 2.5 presents a survey of the annual tritium production rates (inadvertently, deliberately, potential) and inventories in different facility types as well as in nature. The facility types and their worldwide number are given in Table 3.4.

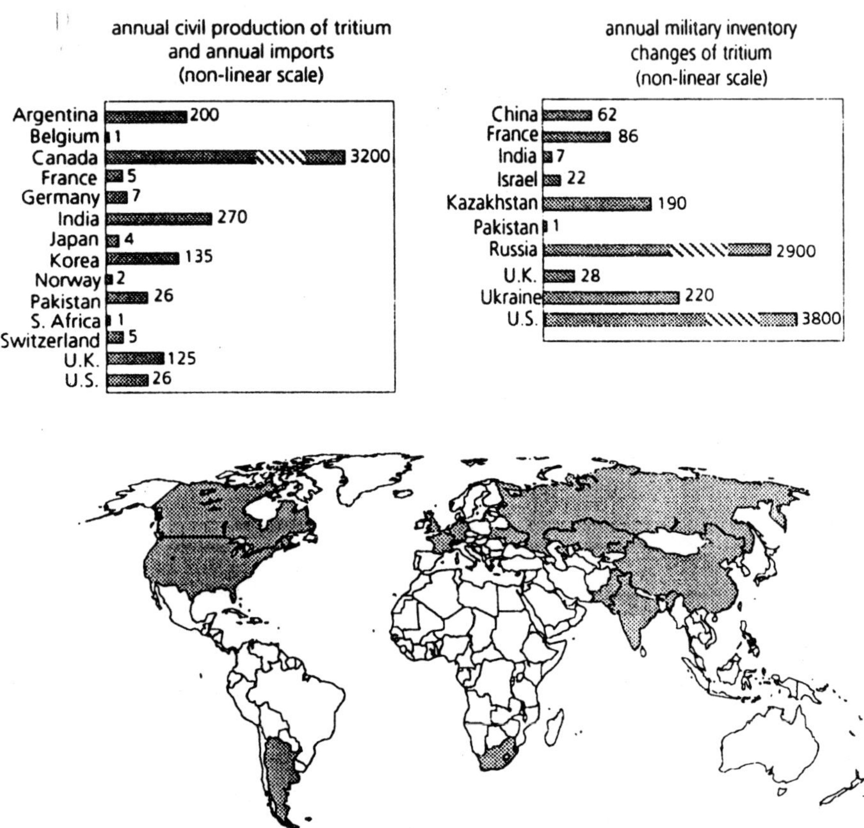

Figure 2.5 World map of annual tritium inventory changes. The diagram has been published in Kalinowski and Colschen (1995). All countries with total changes of more than 1 g/y are highlighted in the figure. Histograms on the left side indicate annual increases in civilian inventories, those on the right military inventory decreases due to radioactive decay. For more explanations see text.

2.8.2 High civilian surplus meets military demand

The distribution of the main inventory changes in civilian and military stocks over the globe are depicted in Figure 2.5. Included are all countries in which either of these inventories changes at a higher rate than 1 g/y.

Histograms on the right side stand for military inventory changes. Since the production rate is in general not known, the column lengths reflect the annual loss due to decay of the assumed military inventories. Civilian production rates and annual imports of tritium are shown on the left with column lengths that correspond to the respective quantities. The largest contribution is made by inadvertent production via the heavy water path, and most of this tritium is not extracted. The fuel rod path is not included in this diagram because its recovery is not of practical significance and because the tritium is typically released during reprocessing at facilities which are in many cases not in the country the fuel comes from.

This world map of inventory changes makes clear at one glance that there are large amounts of surplus tritium produced in civilian facilities in a few countries (Canada, India, Argentina) which are facing large losses of tritium in military stocks due to radioactive decay. Since civilian surplus and military losses do not occur in the same countries, this unequal distribution establishes a potential for horizontal proliferation.

Obviously, tritium supply is a key problem for the maintenance of nuclear weapons for any nuclear weapons state that has no indigenous production capability. It is not possible to buy tritium in quantities which are adequate for nuclear arsenals on the international market and pretend any plausible civilian uses. The whole annual throughput of the civilian world market is more than ten times smaller than the total decay in military stocks.

This calls for a stricter and better coordinated international control of tritium in addition to export control within the framework of the Nuclear Supplier Guidelines which cover tritium and some tritium technology. An international tritium control system which would be comparable to the nuclear safeguards applied to plutonium and highly enriched uranium and which could block or aggravate horizontal proliferation does not yet exist.

2.9 Endnotes

1. A reasoning for these quantities is given in Section 3.2.2.
2. See the description of these facilities in Section 2.6 and the list of all facilities worldwide in Table 3.4.
3. See this section for production paths and the following section for removal paths as well as Table 3.5 for a summary of all acquisition paths.
4. Other principal steps are described in Kalinowski (1989).
5. Short notation: ^{6}Li(n,α)^{3}H; cross-section: $\sigma_{thermal} = 945$ barn, $\sigma_{eff} = 693$ barn (Phillips and Easterly, 1980).
6. Short notation: ^{7}Li(n,nα)^{3}H; cross-section: $\sigma_{fast} = 0.00194$ barn (Tritium Test Facility, 1982), $\sigma_{eff} = 0.0516$ barn (Phillips and Easterly, 1980).
7. Short notation: ^{7}Li(n,2n)^{6}Li; cross-section: $\sigma_{>8MeV} = 0.1$ barn D'Annucci (1982).
8. Since mercuric impurities turn lithium red, the product was given the code-name "red mercury" in the former Soviet Union (Hibbs, 1993). There exist other explanations for red mercury as well.
9. See last columns in Tables 2.1 and 2.2.
10. For example, a lithium rod with an outer diameter of 45 mm in a position with a undepressed thermal flux of 2.6×10^{14}cm^{-2}s^{-1} has a maximum production rate of 5.79×10^{14}cm^{-3}s^{-1} at the target surface and an average production rate of 1.97×10^{14}cm^{-3}s^{-1} (Krug and Weise, 1981). This indicates that calculations may easily be wrong by a factor of 2.

11. If there are more tritium atoms than remaining lithium atoms, tritium can no longer be kept absorbed in the metal. The tritium gas pressure increases and can cause dimensional instabilities, blistering, creep collapse of cladding, and eventually loss of tritium (Church, 1983).

12. According to Church (1983) the discharge burnup is in the range of 400 to 500 $GW_{th}d$, which corresponds to an irradiation cycle of 167 to 208 days at 2400 MW_{th} power. The total core consists of 400 assemblies. The maximum irradiation occurs when about half of the lithium-6 atoms are consumed.

13. The lithium content has to be less than 4%. Otherwise, a separate metallic lithium phase is formed, which would lead to a higher hydrogen content (Abraham, 1963).

14. The $LiAlO_2$ content should be less than 10 vol% of the composition in order to maintain the metallurgical properties of zirconium (Cawley and Trapp, 1984).

15. Two U.S. patents were granted (Benedict *et al.*, 1981).

16. Lu *et al.* (1988) made their calculation for extra plutonium production. As a rough estimate, it is assumed here that the quantity of tritium producible in the same way is 1/80 of that for plutonium.

17. Nevertheless, in Lu *et al.* (1988) a high probability of adverse impact on the safe operation of the reactor is assumed.

18. For the near future, it is not expected that breeding of significant quantities via lithium path for civilian purposes will have to be reported. The only exception might be fusion energy research.

19. This case could be of importance if military tritium production is continued under a fissile material cutoff agreement.

20. In the U.S. this technology is considered as an alternative to build a new production reactor and as a replacement for the military production facility at the Savannah River Plant. Protagonists of this option argue that the time necessary for the development of this technology (at least 5 years) can be saved by implementing the proposed cuts in the nuclear arsenal. Accelerator breeding seems to be a realistic option.

21. 100% capacity assumed (Miller, 1989).

22. The costs were estimated to be $20,000–30,000 per gram (Schriber, 1984).

23. Using a proton accelerator of 1 km length which requires continuous electric power supply of about 740 MW (Krass, 1989).

24. Short notation: $^3He(n,p)^3H$; cross-section: $\sigma_{thermal} = 5327$ barn.

25. Data are given in Phillips and Easterly (1980) with respect to thermal power. An efficiency of 40% is assumed.

26. See Sokolski (1982). In 1982, the Department of Commerce had under consideration a license to export 95 g (710 l) of helium-3 to South Africa. The stated purpose was a rapid power excursion experiment in the Safari-1 test reactor at Penlindaba (tank LWR, 20 MW_{th}, thermal flux: 2.5×10^{14} $s^{-1}cm^{-2}$). The Executive Branch Sub-Group on Nuclear Export Coordination argued that the tritium production rate would be less than 1 g/y due to the low power level of the reactor. Conversion in larger power-producing reactors would face safety and other problems and could therefore be excluded (Thomas, 1982).

27. Much of this information is drawn from Thomas (1982).

28. Short notation: $^{11}B(n,n2\alpha)^{3}H$.

29. Short notation: $^{10}B(n,2\alpha)^{3}H$; cross-section: $\sigma_{fast} = 0.0288$ barn (Tritium Test Facility, 1982), $\sigma_{eff} = 1.27$ barn (Phillips and Easterly, 1980).

30. Short notation: $^{10}B(n,\alpha)^{7}Li$; cross-section: $\sigma_{thermal} = 3837$ barn, $\sigma_{fast} = 10.69$ barn (Tritium Test Facility, 1982), $\sigma_{eff} = 3060$ barn (Phillips and Easterly, 1980).

31. According to Bonka (1980), in absorber rods 0.1 g/(GW_ey) of tritium is produced in HWRs, MAGNOX, and HTRs, 0.2 g/(GW_ey) in BWRs and AGRs. It appears, however, that only the nuclear reaction B-10(n,2α)H-3 was considered.

32. In PWRs and FBRs control rods are made of silver, indium, and cadmium alloy, which does not result in significant tritium quantities. Anyway, they are used only in an emergency. Otherwise, boric acid is used for reactivity control.

33. See D'Annucci (1982). A number of 1436 BPRs was assumed as for the reference PWR in Lu *et al.* (1988).

34. The reference reactor in Lu *et al.* (1988) has a total heat output of 3250 MW_{th}, i.e., about 1 GW_e.

35. The term "inadvertently" is referring to the technical process only. Therefore it is adequate only in the case in which the reactor is built to generate electricity or for research purposes. The choice for an HWR can be made deliberately with the intention to exploit the tritium produced in the heavy water.

36. Short notation: $^{2}H(n,\gamma)^{3}H$; cross-section: $\sigma_{thermal} = 0.57$ mbarn, $\sigma_{eff} = 0.316$ mbarn (Phillips and Easterly, 1980).

37. This reactor is a CANDU type reactor and has a specific production rate of $(60–90)\times 10^{3}$ TBq/(GW_ey), i.e., (170–250) g/(GW_ey) at 100% capacity. See Table 2.3.

38. Commercial LWRs have employed lithium salts which are enriched to 99.9% lithium-7 in order to reduce tritium production (Peterson and Baker, 1985).

39. 0.015% of natural hydrogen is deuterium (Locante and Malinowski, 1973).

40. See Locante and Malinowski (1973). According to Locante and Malinowski (1973), the deuterium alone gives 0.2 TBq/(GW_ey). The lithium used for pH control alone yields 0.3 TBq/(MW_ey) in an equilibrium fuel cycle according to Peterson and Baker (1985). Another study gives an initial value of 0.083 g/(GW_ey). Saturation is reached after three years at about 25 TBq/GW_{th}. This implies an average production rate of 0.07 g/(GW_ey) (Bonka, 1980).

41. See Ebeling (1985). Release estimates from rods with zircaloy claddings range from 0.013% to 1.0%; in case of stainless steel cladding, up to 80% can be released (Phillips and Easterly, 1980).

42. According to IAEA (1981) the range is 0.062 to 0.072 g/t uranium for a burnup of 30 GWd/t; according to Phillips and Easterly (1980), the remaining tritium ranges from 0.03 to 0.08 g/t of uranium; Brücher and Hartmann (1983) give a range of 0.050 to 0.083 g/t of uranium.

43. After this treatment, fuel rods have to be disposed of as waste. It seems not worth reprocessing fuel that has been treated with heat. For the planned German reprocessing facility a voloxidation process was considered as a head-end step prior to dissolution to release the gaseous fission products and tritium. But this was later rejected because indications appeared that above 650°C insoluble PuO_2 is formed, which would increase plutonium losses in the reprocessing stage. The process was intended to run in an oxygen atmosphere at 500 to 700°C (Küsters *et al.*, 1979).

44. Bray (1981) succeeded in releasing 100% of the tritium at 1500°C within 6 hours. Campbell and Pattison (1981) required 24 hours to release 99% of the tritium at 1000°C.

45. Emissions from reprocessing facilities and heavy water emissions from HWRs are covered under the relevant items.

46. Emissions from reprocessing facilities and heavy water emissions from HWRs are covered elsewhere.

47. Total release rate based on critical group impact (Bundesamt für Strahlenschutz (BfS), 1990)

48. Assumptions: 10,000 deuterium shots per year with an average level of 10^{16} deuterium–deuterium fusions per shot (CFFTP, 1988).

49. Fuel fabrication capacities are expressed as mass of heavy metals referring to the actinide elements (uranium, plutonium, etc.) contained in the spent fuel. See Table A.6.

50. At the Hyderabad plant in Pakistan, safeguards apply only to enriched uranium.

51. Spent fuel inventories are expressed as cumulative mass of heavy metals referring to the actinide elements (uranium, plutonium, etc.) contained in the spent fuel (IAEA, 1989).

52. The figures are taken from Table A.9 and multiplied by a factor 0.8 for a crude decay correction and allowing for some 10% of tritium remaining in the waste. The error margin is small for this figure because the type of fuel and its specific tritium content is accounted for.

53. Figures are taken from Table A.9 and are roughly corrected for decay by multiplying by 0.8.

54. Not decay-corrected. One example in the U.S. is the Savannah River Plant. It has a yearly capacity of 2800 t of uranium. Since 1954, some 40,000 t of natural uranium metal have been reprocessed at this facility. Assuming a mean burnup of 600 MWd/tU, about 60 g of tritium have been released.

55. In reality there will always remain the possibility of retrieving the disposed material, but the costs may preclude that.

56. Low inventory: $I < 1$ Ci (< 0.0001 g); medium: $1 \text{ Ci} \leq I \leq 1000$ Ci ($0.0001 \text{ g} \leq I \leq 0.1$ g); high: $I > 1000$ Ci (> 0.1 g) (United Nations Scientific Committee on the Effects of Atomic Radiation (UNSCEAR), 1977).

57. Low inventory: $I < 0.01$ g; medium: $0.01 \text{ g} \leq I \leq 1$ g; high: $I > 1$ g (IAEA, 1991).

58. Tritium stored in single containers has to be included in the accountancy in any case for one of two reasons: Either the place of its location has to be declared, e.g., as a storage facility, and all tritium stored at that place has to be safeguarded; or the container is declared and used for a shipment from one declared and safeguarded location to another facility.

59. An estimation of total tritium in military stocks of the U.S. was made by Cochran (1987). The figure given here is an extrapolation of their estimate taking into account further production up to 1988 and radioactive decay.

60. This decision is described in DOE (1999).

61. This assessment assumes — in contrast to official U.S. estimates (see above) — that the Moscow Treaty will enter into force.

62. Cochran and Norris (1993, p. 154), assumed an annual production rate of 7.3 kg/y until 1984 and 3.4 kg/y in 1985 and 1986.

References

Abraham, B.M. (1963) Tritium Production by Neutron-Irradiation of Aluminium-Lithium Alloys. United States Patent No. 3,100,184, August 6.

Albright, D. and Zamora T. (1989) India, Pakistan's Nuclear Weapons: All Pieces in Place. *The Bulletin of the Atomic Scientists*, 45, No. 5, 20–26.

Arms Control Today (1993) U.S. Tritium Reactor to Remain Closed, May, 27.

Barrilot, B. (1991) *Fabrication des armes nucleaires en France*. Centre de Documentation et de Recherche sur la Paix et les Conflits, Lyon.

Benedict, M., Pigford, Th.H. and Levi, H.W. (1981) *Nuclear Chemical Engineering*, (2nd Edition).

Bergeron, K.D. (2002) *The Tritium on Ice. The Dangerous New Alliance of Nuclear Weapons and Nuclear Power.* The MIT Press, Cambridge, MA.

Bonka, H. (1980) Produktion und Freisetzung von T und C-14 durch Kernwaffenversuche, Testexplosionen und kerntechnischen Anlagen, einschließlich Wiederaufarbeitungsanlagen. In *Strahlenschutzprobleme im Zusammenhang mit der Verwendung von Tritium und C-14 und ihren Verbindungen*, F.-E. Stieve, G. Kistner (eds) STH-Berichte 12/80, Berlin.

Bray, L.A., *et al.* (1981) Thermal Outgasing of Irradiated Fuel, *ANS Transactions*, 39, 219–220.

Brücher, H. and Hartmann, K. (1983) Freisetzung von gasförmigem Tritium aus Wiederaufarbeitungsanlagen. January, Jül-1838, Jülich.

Bruggeman, A., *et al.* (1985) Development of the ELEX Process for Tritium Separation at Processing Plants. *Radioactive Waste Management and the Nuclear Fuel Cycle*, 6, 237.

Bundesamt für Strahlenschutz (BfS) (1990) Plan. Endlager für radioaktive Abfälle Schachtanlage Konrad, Salzgitter. Kurzfassung, Stand 9/86 in der Fassung 4/90, Salzgitter [New emission values according to "Austauschtabelle" dated 30.05.1991].

Campbell, D.O. and Pattison W.L. (1981) Tritium and Fission Product Behavior During Irradiated Fuel Heat Treatments. *ANS Transactions*, 39, 219.

Cawley, W.E. and Trapp T.J. (1984) Lithium Aluminate/Zirkonium Material Useful in the Production of Tritium. United States Patent No. 4,475,948, October 9.

Cawley, W.E. and Trapp T.J. (1985) Fuel Assembly for the Production of Tritium in Light Water Reactors. United States Patent No. 4,526,741, July 2.

CFFTP (1988) Tritium Supply for Near-Term Fusion Devices. CFFTP-G-88024, May.

Church, J.P. (1983) Safety Analysis of Savannah River Production Reactor Operation (deleted version). Savannah River Plant report, DPSTSA-100-1, September.

Cochran, T.B., *et al.* (1987a) *Nuclear Weapons Databook, Vol. 2: U.S. Nuclear Warhead Production.* Cambridge.

Cochran, T.B., *et al.* (1987b) *Nuclear Weapons Databook, Vol. 3: U.S. Nuclear Warhead Facility Profiles.* Cambridge.

Cochran, T.B., *et al.* (1989) *Nuclear Weapons Databook, Vol. 4: Sovjet Union's Nuclear Weapons.* Cambridge.

Cochran, Th.B. and Norris, R.S. (1993) *Russian/Soviet Nuclear Warhead Production*, NWD 93-1, September, Washington.

Commissariat a l'Energy Atomique (CEA) (1962) Rapport annuel.

D'Annucci, F., *et al.* (1982) Tritium Analysis of Irradiated Burnable Poison Rods. *Nuclear Technology*, 59, 9–13.

Dastur, A. and CFFTP/AECL (1986) Internal report. Cited in: CFFTP, *Tritium Supply for Near-Term Fusion Devices*, CFFTP-G-88024, 1988, May.

DeVolpi, A. (1987) *Lithium-cooled LMR for Nuclear-material Production.* Memo dated September 15.

DOE (1999) Fiscal Year 2000 Stockpile Stewardship Plan. Executive Overview. U.S. Department of Energy, Office of Defence Programs, March.

Donnelly, W.H. (1989) Alternative Sources of Tritium for Use in U.S. Nuclear Weapons. A conceptual analysis, prepared for the Honorable Edward Markey, Congressional Research Service, Washington.

Ebeling, N. (1985) Die Verfestigung von angereichertem Tritium aus der Wiederaufarbeitung. Jül-2010, Jülich.

Engholm, B.A. (1983) Monte-Carlo Analysis for the TFTR Lithium Blanket Module. In *Tech. Cont. Nucl. Fusion, Proc. Fifth Topical Meeting*.

Gorman, D.J. and Wong, K.Y. (1979) Environmental Aspects of Tritium from CANDU Station Releases. In *IAEA. Behaviour of Tritium in the Environment*, Proc. Symp. San Francisco, October 16–20, 1978, IAEA-SM-232/49, Vienna.

Governmental Printing Office (GPO) (1982) U.S. policy on export of helium-3 and other nuclear materials and technology. Hearing before the Subcommittee on Energy, Nuclear Proliferation, and Government Processes of the Committee on Governmental Affairs, United States Senate, 97th Congress, May 13, Washington.

Grathwohl, P. (1973) Erzeugung und Freisetzung von Tritium durch Reaktoren und Wiederaufarbeitungsanlagen und die vorraussichtliche radiologische Belastung bis zum Jahre 2000. KfK, externer Bericht 4/73-36, Karlsruhe.

Greenspan, E. and Miley, G.H. (1981) Fissile and synthetic fuel production ability of hybrid reactors. *Atomkernenergie/Kerntechnik*, 38, 12–19.

Gsponer, A. (1984) The French military nuclear fuel cycle. In *Le centrali nucleari e la bomba*, G. Salio (ed.), Report ISRI 82-03, 3rd version December 1983, Ediziori Gruppo Abele, Torino.

Hibbs, M. (1993) "Red Mercury" is lithium-6, Russian weaponsmiths say. *Nucleonics Week*, July 22.

Housiadas, C., Perujo, A. and Vassallo, G. (1994) The Control of Tritium in ETHEL, *Journal of Fusion Energy*, 13, 455–460.

Hugony, P., Sauvage, H. and Roth, E. (1973) La production de tritium en France, *Bulletin d'Information Scientifique et Technique*, No. 178, February, 3–17 [English translation available as *Tritium Production in France*, ERDA-tr-286].

IAEA (1979) Behaviour of Tritium in the Environment. In *Proc. Symp. San Francisco*, October 16–20, 1978, IAEA-SM-232/49, Vienna.

IAEA (1981) Handling of Tritium-Bearing Wastes. Technical Report Series No. 203, Vienna.

IAEA (1989) Nuclear Power and the Fuel Cycle: Status and Trends 1989. Part C of the IAEA Yearbook, Vienna.

IAEA (1991) Safe Handling of Tritium. Review of Data and Experience. Technical Report Series No. 324, Vienna.

Jenks, G.H., *et al.* (1963) Production of Tritium. United States Patent No. 3,079,317, February 26.

Joint Publications Research Service (JPRS) (1988) Selections from China Today: Nuclear Industry. Science and Technology, China. Report JPRS-CST-88-002, January 15, Washington.

Kalinowski, M.B. (1989) Verwendbarkeit und Produktion von Tritium für Kernwaffenprogramme. IANUS-10/1989; Also as appendix in: Müller, H. (1989) Nach den Skandalen. Deutsche Nichtweiterverbreitungspolitik. HSFK-Report No. 5/1989.

Kalinowski, M.B. and Colschen, L.C. (1995) International Control of Tritium to Prevent its Horizontal Proliferation and to Foster Nuclear Disarmament. *Science and Global Security*, Vol. 5, No. 2.

Kobisk, E.H., *et al.* (1989) Tritium-Processing Operations at the Oak Ridge National Laboratory with Emphasis on Safe-Handling Practises. *Nucl. Instr. Methods*, A282, 329–340.

König, L.A. (1980) Produktion und Freisetzung von Tritium und C-14 in der Kernforschungsanlage Karlsruhe. In *Strahlenschutzprobleme im Zusammenhang mitder Verwendung von Tritium und C-14 und ihren Verbindungen*, F.-E. Stieve, G. Kistner (eds), TH-Berichte 12/80, Berlin.

Kouts, H. and Long, J. (1973) Tritium Production in Nuclear Reactors. In *Tritium Conf. Proc.*, A.A. Moghissi, M.W. Carter (eds), August 30 – September 2, 1971, Las Vegas, Nevada.

Kraemer, R., *et al.* (1993) Common Tritium Control Methodology Proposed for Two Civil Tritium Facilities ETHEL and TLK. In *15th Annual Meeting of ESARDA*, May 11–13, Rome.

Krass, A. (1989) The Tritium Problem and the Proton Accelerator. A Report by the Union of Concerned Scientists, September, Cambridge, MA.

Krug, W. and Weise, L. (1981) Berechnung der möglichen Tritium-Erzeugung in Bestrahlungspositionen des FRJ-2 (DIDO) für Fusionsblanket-Experimente. Jül-1727, Jülich.

Küsters, H., LaPovic, M. and Wiese, H.W. (1979) Fuel Handling, Reprocessing, and Waste and Related Nuclear Data Aspects. KfK-2833, Karlsruhe.

Locante, J. and Malinowski, D.D. (1973) *Tritium in Pressurized Water Reactors.* In *Tritium Conf. Proc.*, A.A. Moghissi, M.W. Carter (eds), August 30 – September 2, 1971, Las Vegas, Nevada.

Lu, M.-S., Zhu, R.-B. and Todosow, M. (1988) Unreported Plutonium Production in Light Water Reactors. ISPO-282, TSO-88-1, February, Brookhaven National Laboratory.

Luykx, F. and Fraser, G. (1986) Tritium Releases from Nuclear Power Plants and Nuclear Fuel Reprocessing Plants. *Radiation Protection Dosimetry*, 16, 31.

Miller, J., *et al.* (1988) The CRITIC Irradiation of Li_2O — Tritium Release and Measurement. *Fusion Technology*, 14, 649 (CFFTP-88-23).

Miller, M. (1989) Verification and Safeguards Consideration. In *The Tritium Factor. Tritium's Impact on Nuclear Arms Reduction*, Nuclear Control Institut and The American Academy of Arts and Sciences, Washington.

Moghissi, A.A. and Carter, M.W. (eds) (1973) *Tritium Conf. Proc.*, August 30 – September 2, 1971, Las Vegas, Nevada.

Morrison, P. and Tsipis, K. (1988) Suggested Study of the Production of Tritium in Naval Reactors, Novemver 28, Cambridge, MA.

Müller, W.D. and Hossner, R. (eds) (1990) *Jahrbuch für Atomwirtschaft 1990.* Düsseldorf.

Müller, W.D. and Hossner, R. (eds) (1991) *Jahrbuch für Atomwirtschaft 1991.* Düsseldorf, 1991.

National Resources Defense Council (NRDC) (1989) *Kysthym Complex and Soviet Nuclear Materials Production.* Fact Sheet, Washington.

Neel, K.H. (1978) *Chemie der nuklearen Entsorgung. Teil I.* München.

Norris, R.S. and Arkin, W.M. (1994) U.S. nuclear weapons stockpile. *The Bulletin of the Atomic Scientists*, July/August, 61–63.

Nuclear Control Institut and The American Academy of Arts and Sciences (1989) *The Tritium Factor. Tritium's Impact on Nuclear Arms Reduction.* Washington.

Osborne, R.V. (1979) Hazard and Protection from Tritium Produced in an Experimental Reactor Loop. *Health Physics*, 36, 167–174.

Peterson, H.T. and Baker, D.A. (1985) Tritium production, releases and population doses at nuclear power reactors, *Fusion Technology*, 8, 2544–2550.

Phillips, J.E. and Easterly, C.E. (1980) Sources of Tritium. Oak Ridge National Laboratory, ORNL/TM-6402.

Ragheb, M.M.H. (1981) Implementation considerations of coupling dedicated fissile and fusile production in fusion and fission reactors. *Atomkernenergie/Kerntechnik*, 38, 85–90.

Rupp, A.F., Cox, J.A. and Binford, F.T. (1965) Radioisotope Production in Power Reactors. Oak Ridge National Laboratory, May, ORNL-3792.

Schriber, S.O. (1984) An assessment of accelerator breeding — economics, developments and a staged program, *Atomenergie/Kerntechnik*, 44, 177–180.

Schwarz, H. (1975) Behandlung und Abtrennung der radioaktiven Spaltprodukte Tritium, Edelgase und Jod in Kernbrennstoffwiederaufarbeitungsanlagen. Jül-1223, Jülich.

Smith, J.M. and Gilbert, R.S. (1973) Tritium Experience in Boiling Water Reactors. In *Tritium Conf. Proc.*, A.A. Moghissi, M.W. Carter (eds), August 30 – September 2, 1971, Las Vegas, Nevada.

Sokolski, H.D. (1982) Testimony in hearing before the Subcommittee on Energy, Nuclear Proliferation, and Government Processes of the Committee on Governmental Affairs, United States Senate. In *U.S. policy on export of helium-3 and other nuclear materials and technology.* 97th Congress, May 13, 1982, Governmental Printing Office (GPO), Washington.

Steiner, R.D, *et al.* (1989) He-3 Blankets for Tritium Breeding in Near-Term Fusion Devices. *Fusion Eng. and Design*, 8, 121–125 (International Symposium, April, 1988, Tokyo).

Stern, W.M. (1988) Nuclear Weapons Material Control: Verification of Tritium Production Limitations. Master thesis, MIT, Cambridge.

Stieve, F.-E., Kistner, G. (eds) (1980) *Strahlenschutzprobleme im Zusammenhang mit der Verwendung von Tritium und C-14 und ihren Verbindungen.* STH-Berichte 12/80, Berlin.

Tanase, M., *et al.* (1988) Production of 40 TBq Tritium using neutron-irradiated ^{6}Li-Al alloy. *J. Nucl. Sci. Technol.*, 25, 198.

Thomas, G.F. and Bereton, S.J. (1985) Enhanced Tritium Production for Fusion Reactors via ^{3}He(n,p)^{3}H in the Heavy Water Moderator of a CANDU Reactor, *J. Fusion Energy*, 4, 27–41.

Thomas, H.E. (1982) Testimony in hearing before the Subcommittee on Energy, Nuclear Proliferation, and Government Processes of the Committee on Governmental Affairs, United States Senate. In *U.S. policy on export of helium-3 and other nuclear materials and technology*, 97th Congress, May 13, 1982, Governmental Printing Office (GPO), Washington.

Tritium Test Facility (1982) *Physics Today*, November, 18.

United Nations Scientific Committee on the Effects of Atomic Radiation (UNSCEAR) (1977) *Sources and Effects of Ionizing Radiation*, United Nations.

Ward, L., *et al.* (1981) Confirmation of the photoneutron cross-section for ^{4}He below 33 MeV. *Phys. Rev.*, 24, 317–320.

Willms, S. (2003) Tritium Supply Considerations. In *Fusion Development Paths Workshop*, Los Alamos National Laboratory, 14 January 2003 (http://public.lanl.gov/willms/Presentations/Tritium_Supply_Considerations.pdf).

Wittenberg, L.J., Sanatarius, J.F. and Kulcinski, G.L. (1986) Lunar source of He-3 for commercial fusion power. *Fusion Technology*, 10, 167–78.

Chapter 3

Verification of an international tritium control agreement

3.1 Introduction

As a result of the diversion path analysis (see previous chapter), two different control tasks have to be solved to verify an international tritium control agreement. They are dealt with in this chapter:

Type I: Verify nonproduction. This has to be carried out at declared facilities with inadvertent, actual, or potential production capacities of more than one significant quantity within one inspection period (1 g/y). Verification of nonproduction relies on the nondestructive analysis of possible irradiation targets at nuclear reactors (see Section 3.3.3) to ensure the absence of the raw material for tritium breeding (especially lithium-6). Inspected fuel will be further controlled with containment and surveillance (see Section 3.5). Any tritium produced has to be verified for nonremoval (see type II) after the initial inventory has been determined (see Section 3.4.2).

Type II: Verify nonremoval from existing and declared inventories. This has to be carried out at declared facilities with inventories or throughputs of more than one significant quantity within one inspection period (1 g/y). Verification of nonremoval is achieved by accountancy (see Section 3.4) complemented by containment and surveillance (see Section 3.5) at tritium-handling facilities.

The problem of clandestine facilities is the same as the one faced by the IAEA in connection with the NPT (see Section 3.6).

Before a verification procedure can be outlined, the objectives must first be carefully defined, and the criteria against which success in meeting those objectives can be measured must also be developed (see Section 3.2.1).

3.2 Conceptual framework for verification

3.2.1 Control goals and criteria

For the verification of an international tritium control agreement, concepts and criteria developed by the International Atomic Energy Agency (IAEA) for nuclear safeguards[1] can be transferred and appropriately adopted. The safeguards system that applies for member states of the IAEA is laid down as document INFCIRC/66/Rev.2. The safeguards system that comes into force due to the Nonproliferation Treaty (NPT) was published by the IAEA as document INFCIRC/153 (Corrected) with the Additional Protocol as INFCIRC/540.

The goals for verification depend on the goals of the particular international agreement in question. In general, an international agreement on tritium control would have the goal to prevent the use of tritium originating from declared or undeclared civilian and military sources in nuclear weapons. The main goals of the verification of an international tritium control agreement are detection of and deterrence against noncompliance along with confidence-building by the demonstration of compliance. Further goals are to increase the technical burden for potential proliferators, to clarify uncertainties, and to increase transparency.

These goals are mainly relevant for the parties to the agreement which is assumed to be, and designed as, an internal regime, i.e., controls are applied to members.[2]

The question is, what degree of confidence in the control system is sufficient to achieve the desired goal? With respect to international arms control regimes, effective and adequate verification is distinguished (Scheffran, 1989).

- The objective of "*effective verification*" is developed under the assumption that any potential treaty violation is taken seriously and, therefore, has to be detected with a high degree of probability. Although in nuclear safeguards it is common to speak of "effective" verification, this term is practically understood in the meaning of "adequate" verification defined in most arms control debates. It is admitted that perfect verification cannot be reached in practice and less demanding requirements are applied.

- "*Adequate verification*" can be achieved as long as the desired security goal will not be substantially undermined by undetected cheating, which would be possible due to a detection probability below 100%. The question is whether the safeguards system can be expected to give sufficient confidence to the parties to the treaty that the goal they aimed at when signing the treaty is not endangered. It is assumed that the threat of detection with a certain probability (e.g., 90 to 95%) is sufficient to deter diversion.

  In the case of tritium control, the required verification effectiveness may be even lower than in nuclear safeguards because tritium is not necessary for the production of first-generation nuclear weapons. In particular the timeliness criterion is much less significant as compared to nuclear safeguards (see below).

In Chapter 3, a mainly technical assessment of an international tritium control is given as the conclusion to this study. The following criteria will be applied to the assessment and served as guidelines for the development of control procedures as described in this chapter:

1. **Adequacy and appropriateness:**
   International tritium control should be introduced in a way which is adequate with respect to its military significance. All verification procedures should be appropriate with respect to the detection goals (see above).

2. **Nondiscrimination:**
   International tritium control should be implemented in a way that does not discriminate against any country or group of countries taking part in the control agreement.

3. **Feasibility and completeness:**
   Tritium control should in principle be feasible. All diversion paths which could yield more than one significant quantity of tritium within the desired detection time have to be accessible for control measures. Diversions have to be detectable. The applied measurement techniques and the control procedures should be reliable without loopholes. Cheating or circumventing of control measures should be impossible or at least too expensive or risky to be of practical interest to the potential diverter.

4. **Effectiveness of control:**
   The objective is the efficient, timely detection (detection time t_d) of any diversion of significant quantities (SQ) of tritium from peaceful activities for the production of nuclear weapons or for purposes unknown with a certain detection probability $(1 - \beta)$ and the localization of tritium and satisfactory clarification of anomalies with a low risk of false alarm (probability α). The deterrence of diversion by the risk of early detection is considered adequate. Relevant parameters (SQ, t_d, $1 - \beta$, α) for inspection procedures and criteria of safeguards systems are introduced in the following sections.[3]

5. **Minimum interference with facility operation:**
   Any disturbance of normal operation and any economic burden to the operator of the facility should be minimized. The safeguards approach should use instruments and techniques which do not negatively influence the performance and costs of the observed facility and which do not jeopardize the safety of the plant.

6. **Minimum intrusion:**
   Only information which is indispensable for control purposes should be obtained by inspectors and the controlling authorities. The objective is to protect data and qualitative information that is considered sensitive to national security, protected by patent, or claimed as trade secret.

7. **Synergies with other control procedures:**
   Other control procedures for nonproliferation and radiation protection should not be hampered or compromised by tritium control. Care should be taken to design control procedures for tritium in a way that makes optimum use of nuclear safeguards which are already in place. Possibly nuclear safeguards can be improved by being supplemented with tritium control procedures.

8. **Costs:**
   The costs for a tritium control system should be minimized. Instruments and equipment for efficient measurement and data handling must be available at reasonable costs. The number of inspections in addition to nuclear safeguards should be small.

9. **Effects on civilian uses of tritium:**
   Control procedures should be neutral with respect to civilian tritium uses. They should be the condition for tritium use, but should not encourage or promote it in any way, because the more tritium is circulating the higher is the proliferation risk and the worse the effectiveness of control.

10. **Acceptability:**
    Facility operators and countries should not find themselves at any substantial disadvantage by accepting international tritium controls.

3.2.2 Significant quantities

Significant quantities in nuclear safeguards

In nuclear safeguards the significant quantity (SQ) corresponds roughly to the amount of nuclear material which would be needed for the production of one crude nuclear explosive device.[4] The same amount of material is sufficient for more than one nuclear warhead which makes use of sophisticated compression technology. The significant quantity is larger than one critical mass.[5] This is officially explained by taking into account unavoidable loss that occurs during conversion and manufacturing processes.

Significant quantities for vertical proliferation

The significant quantity for tritium will not necessarily be the quantity required for one single nuclear weapon. The definition will definitely depend on the international agreement in question. A variety of possible agreements can be imagined.[6] The extremes are on one hand an agreement which aims at preventing countries from starting up military-related tritium activities and on the other an agreement among nuclear weapons states which have large quantities of tritium in their arsenals of nuclear weapons. In the latter case, the significant quantity will be in the order of tens or hundreds of gram. In an earlier study the significant quantity for one particular agreement, namely the control of superpower's tritium production, was estimated:[7]

> In the superpower tritium verification regime, the value of a SQ is dependent on existing strategic capabilities. A SQ would correspond to the amount of material required to make a strategically or politically significant difference in superpower arsenals. A SQ with nuclear stockpiles at 20,000 is much different than a SQ with nuclear stockpiles at 1000 weapons.
>
> Moreover, the definitions of SQ depends on the generic goal of the treaty as well as the assumptions of the regime. If a clandestine stockpile of tritium is assumed a priori, then a SQ would correspond to that amount of tritium required simply to *sustain* a significant number of weapons (i.e., 5.5% of significant number of weapons).

> In the case of 12,500 weapons, this analysis assumes that a SQ would correspond to sustaining 500 excess weapons. Thus, an annual SQ would be approximately 100 grams.[8] In the assumed case of stockpiles required for minimum deterrence, a SQ is assumed to correspond to sustaining 200 excess weapons or approximately 40 gram of excess production per year.[9]

After implementation of the Moscow Treaty with 1700 to 2200 strategic nuclear warheads remaining for Russia and the U.S., 200 excess weapons might mark a significant break-out. Their maintenance would require at least some 22 g/y assuming two grams are required for each warhead (see Section 1.3.2). Thus, in case of an agreement between Russia and the U.S. on a verified cutoff for the production of fissionable materials and tritium ("integrated cutoff")[10] the SQ might be set at 20 g for tritium.

Significant quantities for horizontal proliferation

For nuclear threshold states with small numbers of nuclear weapons, even a single boosted nuclear warhead can make a difference, since the yield of a warhead can be boosted by a factor of between two and ten. Therefore, an SQ may be proposed that is aimed against the horizontal proliferation of one boosted nuclear weapon. The average amount of tritium estimated to be used for boosting is between two and three grams. Original information concerning the minimum quantity of tritium necessary for a boosted or thermonuclear warhead is classified. Any state that intends to use tritium for weapons can make use of less tritium for research purposes and to gain experience in tritium handling. For neutron generators much smaller quantities are required, but this is not considered to be of high significance because technical alternatives working without tritium are available.

In this chapter, it is assumed that an SQ is one gram of tritium. This is more conservative than in nuclear safeguards, and it can be expected that the SQ will not be smaller in any international agreement on tritium control. It may in practice be set at, for example, 10 g. If it can be shown that verification is feasible with such a conservative SQ, it would work even better for any realistic value of the SQ.

In Table 3.1 the quantities of world plutonium and tritium stocks are compared as of 1993. Though the total number of significant quantities existing in the world's civilian and military stocks is nearly equal for both materials (around 150,000), the civilian inventory is much higher for plutonium (110,000) than for tritium (25,000). Most challenging is the separated material at bulk-handling facilities. At the end of 1993, there were some 23,000 significant quantities of separated civilian plutonium in the world, which is nearly three times as much as significant quantities of separated civilian tritium (8000). By 2003 civilian inventories in separated form increased for both plutonium and tritium. The military inventories of tritium decreased at least in one nuclear weapons state. Therefore, qualitatively, the situation has not changed. This comparison gives a first indication for a positive answer to the question of whether the verification of an international tritium control is feasible.

Significant quantities for economic and radiation protection purposes

Tritium accountancy is not only of interest for nonproliferation but also for economic reasons and for the purpose of radiation protection. Thus, it is interesting to note

Table 3.1 Comparison of world inventories of significant quantities and expected accountancy capabilities for plutonium and tritium. Source: Colschen and Kalinowski, 1994.

	plutonium		tritium	
Significant quantity SQ [a]	8 kg		1 g	
δ_{MUF} at bulk-handling facilities (sect. 3.4.3)	0.005–0.010		0.0025–0.050	
quantities	in t	in SQ_{Pu}	in kg	in SQ_T
total world civilian inventory, end of 1993	890	111,000	25	25,000
subject to IAEA safeguards, end of 1992	404	50,500	0	0
separated world civilian inventory, end of 1993	180	22,500	8	8,000
total world military inventory in 1993	250	31,250	140	130,000
expected accountancy capability E (sect. 3.4.3)				
for separated world civilian inventory	2.96–5.92	370–740	0.066–1.32	66–1300

[a] It should be noted that the definition of SQ_T for tritium is more conservative than it is for plutonium. SQ_{Pu} is larger by a factor of about 2 than the actual average amount of plutonium assumed to be used in nuclear warheads, whereas SQ_T is smaller by a factor of 2 to 3 than the average quantity of tritium believed to be inserted in weapons. Therefore, the assessment for tritium is more conservative than it is for plutonium by a factor of about 4 to 6.

that quantities which might be considered significant for economic or radiological purposes would be lower than the weapons-related SQ. A loss of 1000 TBq (2.8 g) of tritium would cost at least ~\$80,000. With respect to radiation protection the loss of any quantity is dangerous, but whether a facility is out of compliance with radiation protection regulations depends on the kind of emission. A possible case would be a sudden release of 0.1 mg of tritium inside a building, which would under certain plausible circumstances be enough for an employee to raise the inhaled amount above the permitted dose limit.[11]

Accountable quantity

The U.S. Department of Energy (DOE) has a system of accounting and control. All quantities of tritium equal to or greater than 0.0005 g (0.18 TBq) are considered accountable within one material balance area (MBA); smaller amounts can be neglected and rounded to zero.[12] For DOE reporting purposes the accountable quantity of tritium for each facility is one order of magnitude larger (Wall and Cruz, 1985).

3.2.3 Relevant time frames

The detection time t_d is the time which elapses between a tritium diversion and its detection. It includes production and even preparations for production, if these steps are illegal and can be detected by the control procedures. The conversion time t_c is the time required to process diverted tritium before it can be inserted into nuclear weapons. The length of the conversion time depends on the availability and also the chemical and physical form of the diverted tritium, i.e., tritium in elemental and gaseous form has a shorter conversion time than oxidized tritium in tritiated water. It is by IAEA standards a determining factor regarding the frequency of inspections, i.e., the inspection period t_i.

The goal of timely detection is defined by the ability to detect a diversion before the diverted material can be used to manufacture a nuclear weapon,[13] i.e., $t_d \leq t_c$. This condition is fulfilled by choosing inspection periods which are shorter than the conversion time ($t_i \leq t_c$).

If the timeliness goal of nuclear safeguards were applied to tritium control in the same manner, inspection frequencies of higher than 52 times per year may be required, since the minimum period for preparing diverted tritium for use in warheads could be as short as a few days.[14] This inspection frequency would not be practical in terms of required manpower, but it is also out of proportion, because of the limited significance of tritium for nuclear weapons in relation to fissile materials.

With respect to tritium accountancy for horizontal nonproliferation it is recommended here that inspections should be performed at least annually, i.e., $t_d \leq 1$ y. This recommendation cannot be derived from technical considerations, but appears plausible as a natural time frame which matches well with other periodic activities in the inspected facilities. Therefore, it is reasonable to work with $t_i = 1$ y as a guideline and adjust the actual inspection periods to coincide with nuclear safeguards inspections or periodic operator breaks for maintenance or other purposes. As far as reactors are concerned, the reasonable solution would be to tie tritium inspections to the routine IAEA inspection procedures for plutonium and HEU.

To avoid the sudden appearance of large accumulated quantities of diverted tritium and to assure a more timely detection of any significant anomaly, it might be necessary to aim at a more frequent accountancy, especially at large tritium gas-handling facilities. But more frequent accountancy typically implies reducing the detection probability because each measurement is associated with an error which propagates into the closure of the material balance. This is another optimization problem.

With respect to vertical proliferation, the desired detection time may be different. In a superpower's tritium verification agreement, the definition of timely detection depends on the strategic capabilities. In the case of a verified limit on tritium production by the superpowers, the insertion of quantities of lithium sufficient to produce tritium in excess of agreed limits would suggest an intention to violate the agreement. If 100 g were considered to be one significant quantity, routine inspections with quantitative assessment of lithium targets inserted in the reactor core would have to be carried out weekly.[15]

The restrictions with respect to the timeliness goal bring up the question of "efficiency." This requirement can still be satisfied. First, although the conversion time may be as short as a few days, it is highly unlikely that the first diverted significant quantity of tritium can immediately be used in a nuclear weapon without previous activities to establish because the handling of tritium requires special precautions and dedicated technologies as well as facilities. Second, the inspections would still fulfill a veritable deterrence function. If the verification procedures are politically satisfying and acceptable, there will be no need to stretch technical verification goals beyond this required level.

Just for comparison, it should be noted that not only are the quantities of tritium which are significant for radiation protection much smaller than for nonproliferation, but this is also true for the required detection times. Radiation protection requires instantaneous (real time) detection of a significant tritium loss. This proves that verification of tritium nonremoval is technically feasible.

3.3 Verification of nonproduction

3.3.1 Verification of production limitations

This kind of verification becomes relevant only if production of tritium is allowed below a certain limit. This may be the case when a limited production for military purposes was agreed upon or when tritium is produced during a peaceful experiment in fusion energy. The verification would be based on accountancy of the raw material (typically lithium-6), taking of an initial physical inventory of the amount of tritium produced (see Section 3.4.2), and subsequently verifying the nonremoval of tritium (see Section 3.4).

An appropriate verification scheme was outlined earlier in the context of the verified control of nuclear weapons materials production by the U.S. and Russia. The assumption was that tritium is produced via the lithium path and that a limit for the maximum production rate has to be verified. According to that study, the following inspection activities are required (Stern, 1988, p. 83):

1. inspector surveillance during fuel reload
2. continuous optical camera surveillance during reactor operation
3. item count and seal application at fuel and target fabrication facility
4. measurement of tritium in extraction tank
5. measurement of lithium-6 content prior and after irradiation to determine burnup of targets
6. application of seals to ensure only measured target assemblies are placed in core
7. sample waste and assess tritium content

In this verification approach the accountancy of lithium-6 plays a key role.[16] "Insertion of quantities of lithium sufficient to produce tritium in excess of agreed limits would suggest an intention to violate the agreement." In addition, too high a target burnup can indicate that an excess amount of tritium has actually been produced in the target.

By measuring the lithium-6 content before and after irradiation, the total amount of tritium produced can be determined. In each case three quantities have to be determined by measurement or calculation (Stern, 1988):

1. total weight of target
2. weight fraction of lithium in target material
3. enrichment in lithium-6

Methods and limitations to determining the quantity of tritium produced are discussed in the section on uncertainties in determining the baseline of tritium inventories (see Section 3.4.2).

Though it was claimed that a tritium production limitation would be verifiable, it is evident that a ban on tritium production would be much easier and less intrusive to verify.

A special variety of production limitations would impose no limit to tritium production, but would completely prohibit plutonium production in the tritium production facility. This is likely to become relevant when tritium production is continued under a verified fissile material cutoff agreement. The nonproduction of plutonium can only be verified by checking that tritium production is not abused to cover up plutonium production. This is because tritium and plutonium production are in competition with each other in consuming neutrons generated in the production reactor (or accelerator). This verification task would be much more intrusive than verification of an integrated cutoff including tritium and it would provide insight into the tritium production rate.

To begin with, all activities of tritium production would have to be declared. Even without access to the tritium production facility, first indications of illegal plutonium production could be seen from the reactor operating cycles. The reactor would have to be shut down to reload fuel and targets. In order to get supergrade plutonium, targets are typically irradiated for 30 days, and a period of 60 days is used for weapons-grade plutonium (6% ^{240}Pu). Tritium production typically requires longer production cycles of some 200 days. After a cycle of this length, the quality of the plutonium isotope composition would not satisfy the standards of a nuclear weapon state.

More adequate verification requires access to the production reactor and intrusive control measures have to be performed. Inspections are required at least during the phase of reloading the core. All fresh fuel and target elements would have to be nondestructively checked for natural or depleted uranium, the raw material used for plutonium breeding, by determining the uranium content and enrichment using neutron coincidence counters. These targets would have to be safeguarded by containment and surveillance so as to verify that they are not reprocessed to extract plutonium. All fuel and target elements would have to be tagged to ensure that they can be identified again at the next shutdown.

During the operating cycles, the presence of inspectors would not be required. The charge and discharge machine as well as the access to fuel and target positions in the reactor core would be sealed. Video cameras would be installed to survey the relevant areas inside the reactor building. Seals and video tapes would be examined at regular time intervals to verify that no changes of the reactor core have occurred in the meantime.

In order to verify that all neutrons are used for tritium production and none are left for covert plutonium production, inspectors would need sufficient information to assess the rate of tritium production. This information has always been considered highly critical for national security in all nuclear weapons states.

3.3.2 Verification of inactivity of production facilities

If nuclear weapons states agreed on a ban on the production of fissile material and tritium for nuclear weapons purposes, there would be the need for verification with inspections which are limited to avoid unwarranted intrusion and proliferation risks. Within the context of proposals towards a fissile material cutoff agreement, verification questions have been discussed,[17] which are to a large extent based on remote sensing.[18] The detection of possible clandestine production facilities is a sensitive issue within an appropriate treaty (see Section 3.6).

The role of tritium with respect to verifying the inactivity of production facilities depends on its inclusion in a cutoff agreement. Two cases can be distinguished:

Case 1. The easiest way of verifying nonproduction of tritium and plutonium is to shut down and dismantle the production reactors. The parties to such an agreement would exchange the relevant data of their tritium production facilities, especially the location of production reactors. The operational status (operating, standby, cold standby, dismantled) of the declared military production facility could be verified by remote sensing using national technical means or by a future international satellite verification agency, possibly under IAEA or UN auspices.

Case 2. If military facilities for the production of tritium are shut down but not dismantled, they have to be kept in a status which is comparable for all states participating in the agreement. The status proposed here is called "cold standby." There is no internationally recognized definition of this status in which a reactor needs preparations of at least a few months to be restarted. Therefore, a definition and related verification measures have to be agreed upon. This may include, for example, the following provisions:

1. fuel and moderator are removed from the site
2. only the minimum staff and systems required for maintaining the infrastructure are kept operating
3. regular safety checks may be continued and a checklist is set up for repairs in case of a restart
4. some parts of the facility may be placed in a more permanent layup mode, e.g., doors sealed, pipes closed

The nonproduction of tritium or plutonium can be verified simply by observing the absence of heat generation, which indicates the inactivity of the production reactors. It is well known from various studies that the technological capabilities have progressed sufficiently and that this verification is feasible.[19] Another valuable advantage of these verifications is the possibility of avoiding inspections that would imply unwarranted intrusion and proliferation risks. The other characteristics of "cold standby" can be verified by unintrusive on-site inspections.

3.3.3 Detection of breeding activities

An analysis of present IAEA nuclear safeguards for plutonium and highly enriched uranium shows that they could be extended to meet tritium control demands. Most activities which would have to be carried out by inspectors at nuclear installations to verify the nonproduction of tritium are already covered by routine nuclear safeguards procedures. In general, safeguards designed to detect clandestine production of plutonium in declared facilities would detect production of tritium as well, requiring only a few additional measures.

Most scenarios of unreported breeding of tritium, e.g., from lithium-6, can be detected by those safeguards activities which are implemented in order to detect unreported breeding of plutonium from natural uranium. The relevant criterion for implementing safeguards is the capability to detect diversion of one significant quantity of plutonium, which is set by the IAEA at 8 kg. Since tritium production is always complementary to plutonium production, those production paths which can produce more than 110 g, the tritium equivalence of 8 kg plutonium, are covered by this criterion and thus already under nuclear safeguards.

All diversion activities related to the insertion or removal of targets for tritium breeding into the reactor core can be detected by those inspection procedures which are designed to detect the insertion or removal of targets which can be used to breed plutonium. For example, lithium target detection in power reactor fuel can be achieved by nondestructive safeguards applied by the IAEA at the fuel fabrication facility. After inspection, the fuel assemblies are sealed. Further safeguards are carried out by item counting, seal inspection, containment, surveillance, and some specific measures as described in Section 3.7.

Instruments used by the IAEA for physical inventory verification of fresh fuel are the hand-held assay probe (HM-4), a portable mini multi-channel analyzer (PMCN), and neutron coincidence counters. The HM-4 and PMCN give a semi-quantitative measurement of total uranium and enrichment. They can detect non-uranium rods in outer rows of the fuel assembly. It is not clear, however, whether undeclared lithium-6 rods located in the interior of an assembly could be detected with these instruments.

The main IAEA safeguards instrument for physical inventory verification of fresh fuel elements is the Neutron Coincidence Collar (NCC).[20] It provides a semi-quantitative measurement of the enrichment level and the total uranium content. It can detect non-uranium rods in the whole fuel assembly. The production scheme most difficult to detect would be the covering up of lithium in the fuel by declaring it to be gadolinium. Monte Carlo calculations indicate that even under such unfavorable circumstances routine measurements with the Neutron Coincidence Collar would result in anomalies (Kalinowski, 1997). If these cannot be resolved, further, possibly destructive, investigations would be triggered.

For nondestructive identification of lithium-6, active interrogation with neutrons or gamma rays is necessary. A dozen different measurement principles have been investigated by Kalinowski (1997). At least one of them appears promising: nuclear resonance fluorescence (NRF) making use of the 3.56 MeV level of lithium-6. The target containing lithium-6 would be irradiated with a continuous Bremsstrahlung source having an end-point energy of 5 MeV. At a right angle, a gamma detector would reveal a peak at 3.56 MeV due to resonance fluorescence, if enough lithium-6 is exposed to the gamma beam. Extrapolations from measurements at the S-DALINAC demonstrate that under a realistic scenario, a measurement time in the order of 10 minutes would suffice to detect 10 g of lithium-6 in one target rod having a linear concentration of 30 mg/cm. The required instrumentation is portable and suitable for routine in-field inspection. The interrogating gamma beam does not pose an unacceptable radiation hazard to personnel and does not cause significant activation or collateral damage to the inspected specimen (Kalinowski, 1997).

Another method which appears less promising is gamma-ray spectroscopy used to identify peaks from the ^{6}Li(n,γ)^{7}Li reaction at 0.478, 6.78, and 7.26 MeV. Extrapolations from experiments demonstrate that in the order of eight hours' gamma spectrum acquisition time would be required to detect 10 g of lithium-6. The assumed neutron source strength of 10^8 neutrons per second is the maximum tolerable under radiation protection guidelines for nuclear safeguards inspections (Kalinowski, 1997).

Spent fuel is investigated with Cherenkov viewers and neutron and gamma measuring devices like the ION-1/Fork. The Cherenkov Viewing Device (night vision device) is used to detect gross defects in spent fuel assemblies. Missing rods can be detected by an additional spot of Cherenkov glow even when they are replaced by dummy rods. This can indicate unreported breeding of plutonium as well as of tritium. The method is not reliable if only a low proportion of rods is removed.

Weighing of fuel assemblies is not yet part of IAEA inspection activities. It would, however, be an easy method to indicate substitution of nuclear fuel by lithium-6 because the weight would be substantially reduced. One fuel rod weighs some 2.5 to 3 kg. If fuel is completely replaced by a lithium-containing compound with density in the order of 2 g/cm^3, the rod content weighs about 300 g and the cladding about 100 g. Hence, the accuracy necessary to detect that even a single fuel had been replaced by a lithium-containing rod is well within the scope of various weighing devices (Lu, *et al.*, 1988). Weighing would provide additional information to detect unreported production of plutonium and would thus improve the efficiency of nuclear safeguards.

3.3.4 Verification of inadvertent tritium production

Large amounts of tritium are produced inadvertently during normal operation of nuclear reactors, especially via the ternary fission path (all reactors) or via the tritiated water path (heavy water reactors only). Since these production paths cannot be banned by an agreement on international tritium control, the amount of tritium has first to be assessed (see Section 3.4.2) and then the nonremoval of the material produced has to be verified (see next section).

In the case of the ternary fission path, nonremoval can be verified by containment and surveillance of spent fuel as it is current practice until the fuel is either reprocessed or safeguards are terminated due to final disposal. At large reprocessing facilities, tritium accountancy would have to be applied. In the case of tritiated heavy water, tritium controls could benefit from the introduction of containment, surveillance, and accountancy of heavy water.[21] These measures would have to be supplemented by determining the concentration of tritium in the heavy water.

3.4 Verification of nonremoval

3.4.1 Methodological background of tritium accountancy

Methods of material accountancy and safeguards systems analysis are frequently applied to fissionable materials (see especially Avenhaus [1977] and Avenhaus [1986]). Theoretical considerations based on these methods led to a number of general results which can be applied to other control systems which are based on material accountancy. One result is that inspectors should ideally invest at least the same measurement effort as the operator does to achieve a material balance. This is not achievable with the limited available inspection resources and would be a large burden for the operator. In practice, the number of verification measurements will be dictated by available manpower resources, material accessibility, plant safety regulations, available measuring devices and their accuracy, etc. From these factors and in-field experience, a pragmatic and realistic inspection goal should be determined.

To assure the continuity of knowledge about the location of tritium under a compromised inspection program, the proposed safeguards scheme is a combination of accountancy (closing a material balance) supplemented by containment and surveillance (see Section 3.5). Material balance closing is based on two different approaches: *attribute and variable verification.* The verification methods should be combined to allow both variable and attribute verification of tritium inventories during physical inventory verification (PIVs). They are also used during interim inspections to ensure continuous operation of the facility and to verify inventory changes.

Variable verification. A statistical test concerning a specific property of a product or process, based on the measurement of the characteristic of one or more samples. A simple example of this type of test is the comparison of the arithmetical mean of a set of measurements of tritium concentration in a sample of material with a similar set of measurements performed independently (IAEA, 1987a).

The goal is to produce quantitative estimates of the different components of the material balance, e.g., PVT/concentration determinations. This verification method is mainly dealt with here.

Attribute verification. A combination of records auditing, comparison of records with reports, and other methods. This is typically done as a statistical test. One example for an attribute test in nuclear safeguards is the measurement of the attribute "radiation emission" to detect dummies among spent fuel elements. These are mostly standard procedures of nuclear safeguards and they are therefore dealt with only marginally in this study.

Material balance area (MBA)

A complete tritium-handling facility may be defined as one single material balance area (MBA) with the walls of the surrounding building being the outer border of the MBA. Alternatively, one facility might be separated into several MBAs. Single experiments or parts of equipment (e.g., tritium storage containers) might be defined

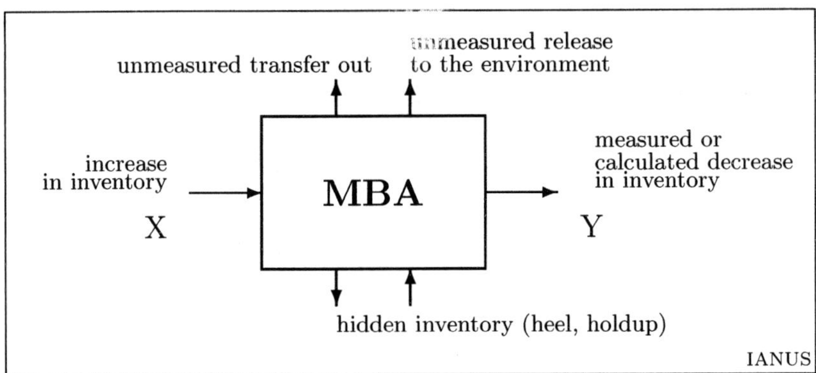

Figure 3.1 Material balance area (MBA) with inputs and outputs.

as one MBA. Inventory balances are closed for each MBA, and any transfer from one MBA to another has to be recorded and verified by measurements.

Dividing a facility into more balance areas will make it easier to localize assumed anomalies. But theoretical investigation shows that with more balance areas the accountancy effectiveness decreases because the errors of each single measurement contributes to the overall error. This indicates an optimization problem.

Material unaccounted for (MUF)

A physical inventory verification (PIV) of all tritium in each material balance area (MBA) of a tritium-handling facility is conducted at regular time intervals. In the interim period, accountancy is made by adding increases in inventory (X)[22] to and subtracting decreases in inventory (Y)[23] from the last physical (or real) inventory I_0 to give the book inventory B:

$$B = I_0 + X - Y \tag{3.1}$$

This is the amount of tritium which should be in the material balance area. The book inventory B at the end of an inventory period is then compared to the new physical inventory I_1. The difference is the material unaccounted for (MUF):[24]

$$\text{MUF} = B - I_1 = I_0 + X - Y - I_1 \tag{3.2}$$

Apart from any undeclared removals D, which should be zero, there are three kinds of contributions to MUF:

1. The instrumental measurement error E_1[25] occurring in the inventory period in question causes random and systematic contributions to MUF. This relates to tritium, which is available for measurements.
2. An unaccounted process loss L causes a systematic contribution to MUF. This is tritium, which is irrecoverably unavailable for measurements.
3. A change of the hidden inventory ΔH causes a temporary contribution to MUF. This is tritium which is recoverably unavailable for measurements.

$$\text{MUF} = D + E_1 + L + \Delta H \tag{3.3}$$

Differences between measured and book values should be within the limits of the measurement error which is associated with closing a material balance. Due to the uncertainty of measurements and due to a portion of tritium not being available, criteria have to be developed which enable a decision as to whether the discrepancy is too large ("significant") and must therefore be regarded as an anomaly and further investigated.

Test of significance

The test of significance decides whether a nonzero MUF could be explained by a measurement error or whether it could be due to diversion or loss.

The null hypothesis H_0 assumes that no material was lost or diverted, i.e., D and L are zero and only the measurement error E_1 and the change in hidden inventory ΔH are responsible for MUF. The alternative hypothesis H_1 assumes that the amount of material M_1 has been lost (L) or diverted (D) with M_1 being equal to MUF.

Since no measurement can be made without random and systematic errors, the hypotheses H_0 and H_1 are true with a certain probability. A test of significance has to be performed in order to decide whether a nonzero MUF can be explained by measurement errors and a hidden inventory (H_0), or diversion or loss should be assumed. For this purpose, a threshold S[26] is chosen. H_0 is taken to be true if MUF $< S$; H_1 is taken to be true if MUF $> S$. Obviously both decisions can be wrong with a certain probability. It is called the *false alarm probability* α when the second decision is wrong; β is the probability that the first decision is wrong. Hence $(1 - \beta)$ is the *detection probability.*

Under the assumption of normally and independently distributed measurement errors (including ΔH) and random losses, there exists a quantitative relationship between α, S and $\sigma(\text{MUF})$:

$$1 - \alpha = \Phi\left(\frac{S}{\sigma(\text{MUF})}\right)$$

In combination with the amount M assumed to be diverted, the detection probability $(1 - \beta)$ can also be calculated:

$$1 - \beta = \Phi\left(\frac{M}{\sigma(\text{MUF})} - U_{1-\alpha}\right)$$

with

$$\Phi(x) = \frac{1}{\sqrt{2\pi}} \int_{-\infty}^{x} exp\left(-\frac{t^2}{2}\right) dt$$

being the standard normal distribution function and U its inverse (see, e.g., Avenhaus, 1986, p. 350).

In order to determine inspection goals and plans, the detection probability $(1 - \beta)$ can be preselected as a basic parameter. International working groups have concluded that in nuclear accountancy β should be low (0.1–0.01). For planning purposes, $(1-\beta)$ is normally set at 0.90–0.95, α is set at 0.05 or less (IAEA, 1987a). The thorough evaluation or investigation of observed anomalies or discrepancies results in a false alarm probability in nuclear safeguards, which is far below this value.

This method was applied to simulate an idealized process in the Tritium Laboratory Karlsruhe (TLK) to study the effects of inspection frequency and waste quantities on accountancy capability (Avenhaus and Spannagel, 1988). In order to demonstrate the false alarm probability and the accountancy efficiency, Gabowitsch and Spannagel (1989) consider the case of a one balance area laboratory in which a 1 g batch of tritium is handled. They assume that the measurements could be performed with a coefficient of variance of $\simeq$1%. They show that the accountancy varies as a function of tritium unaccounted with $\alpha = 1$, 5, and 10% as parameters. As expected, accepting an increased false alarm probability leads to an increased accountancy effectiveness.

Measurement accuracy and expected accountancy capability

The technical problems in accountancy can be aggregated and compared by defining the *expected accountancy capability* E, which is the minimum loss of nuclear material which can be expected to be detected by material accountancy. The accountancy capability is part of the technical capability, which is a measure for the expected performance of a system of accountancy measures. It depends on the measurement uncertainty expected when closing a material balance. This relationship can be quantified to a certain extent. The following algorithm is used by the IAEA.[27]

A quantity E is defined by the equation

$$E = 3.29\ \delta_E\ A \qquad (3.4)$$

where A is the amount of material in the material balance expressed as the larger of the inventory or throughput, the factor 3.29 corresponds to a detection probability of 0.95, and a false alarm probability of 0.05; δ_E is the *expected measurement accuracy* for closing a material balance (i.e., the expected accuracy of MUF) (IAEA, 1987a).

Expected measurement accuracies in nuclear safeguards based on international standards of accountancy, i.e., considered achievable in practice at bulk nuclear facilities, range from 0.002 for uranium enrichment plants and 0.01 for plutonium reprocessing plants to 0.25 for separate waste storages (IAEA, 1987a). The theoretical value for heavy water in power reactors is $\delta_E = 0.005$ [28] For comparison, expected measurement accuracies in nuclear safeguards are presented in Table 3.2.

In nuclear as well as in tritium accountancy $\delta_E\,A$ is not only due to uncertainties in the measurement procedure itself—E_1—(accuracy of used measurement instruments, accuracy of standards and calibration, sampling from inhomogeneous bulk material, loss and impurities in sample preparation, etc.), but there is also a considerable contribution—$L + \Delta H$—from the fact that there is an uncertainty in the quantity of tritium that is freely available for measurement. This is due to losses and recovery from hold-ups that occur before the measurements are made.

Table 3.2 Expected measurement accuracy δ_E (relative standard deviation) associated with closing a nuclear material balance in nuclear safeguards.

bulk facility type	δ_E	ref.
uranium enrichment	0.002	(IAEA, 1987a)
uranium fabrication	0.003	(IAEA, 1987a)
plutonium fabrication	0.005	(IAEA, 1987a)
uranium reprocessing	0.008	(IAEA, 1987a)
plutonium reprocessing	0.010	(IAEA, 1987a)
separate scrap storage	0.04	(IAEA, 1987a)
separate waste storage	0.25	(IAEA, 1987a)
heavy water in power reactor	0.005	(Morsy, 1987)

Although it may not be easy to separate both contributions to δ_E in all cases, to do so will still promote understanding of the technical problems with tritium accountancy.

$$\delta_E \, A = E_1 + L + \Delta H$$

From the objectives and criteria outlined in this section, it follows that four tasks have to be solved with respect to tritium accountancy. They are all covered in the following sections:

- Determine the baseline: How can the initial inventory be taken after production of fresh tritium? (see Section 3.4.2)
- Investigate the instrumental measurement accuracy E_1: How well can available tritium be measured? (see Section 3.4.3)
- Analyze mechanisms of loosing tritium: How much tritium is unavailable for inventory taking due to unaccounted losses? (see Section 3.4.4)
- Analyze mechanisms of holding back tritium: How much tritium is unavailable for inventory taking due to hidden inventories? (see Section 3.4.4)

3.4.2 Uncertainty in baseline determination

The uncertainty in the baseline is revealed by comparing the initial physical inventory to quantities of tritium as derived by calculations. This uncertainty may be due to wrong inventory measurements, to incorrectly quantified transfers, and in the special case of tritium production it may be due to uncertainty in predicting the tritium production rate. Accurate accountancy of tritium depends on precise knowledge of the production rate. It seems not to be feasible with an appropriate effort to calculate the production rate with the required accuracy, because too many varying parameters play a role.

The case of fusion reactors would be the most difficult with respect to baseline determination, if they were ever realized. A lot of work has been done to calculate tritium breeding rates in fusion fuel cycles. For example, Gabowitsch and Spannagel developed the computer model KATRIM (Karlsruhe Tritium Model)

(Gabowitsch and Spannagel, 1989) for this purpose. Their calculation of the tritium inventory in the blanket processing unit of the UWMAK design[29] showed a considerable variance with time, which is large for plant availabilities (capacity factors) around 50% (dynamic case) and small for availabilities close to 0 or 100% (stationary case). In the case of 50% the inventory fluctuates between 0.3 and 1.0 kg due to several scheduled or random shutdowns and restarts. As a consequence, the tritium inventories of other subsystems in contact with the blanket will fluctuate as well. For example, at an availability of 30% the inventory of the blanket coolant increases from 0 to 40 g within 400 days of operation and fluctuates with more than ±5 g. Though the momentary inventory is known with an accuracy smaller than the fluctuations, in the very likely dynamic case of reactor operation, the inventories will be known with a considerable uncertainty.

In another case, calculations with respect to the Jülich research reactor FRJ-2 (DIDO) in-pile experiments have been carried out to determine the tritium production rates under conditions characteristic for fusion reactor blankets (Weise, 1986). Error propagation calculations showed that the microscopic production rate could be determined with an accuracy of ±13.4% ($1\,\sigma$).

The rate of military tritium production at Marcoule (France) is calculated from the ratio of consumed lithium-6 to initial lithium-6 rather than determined by theoretically calculating the neutron irradiation rate of the lithium target (Hugony *et al.*, 1973).

Tanase *et al.* (1988) determined the amount of tritium in small ^{6}Li-Al alloy targets prior to extraction by calorimetry and by measuring volumetrically the amount of helium (both ^{4}He and ^{3}He) which was released in the extraction step. The contents were in the range of 1–2 TBq per target and 40 TBq (0.1 g) in whole batches. They also reported earlier measurements with 4 TBq batches. Although no accuracies of measurements were given, uncertainties can be estimated from the deviations between calorimetry and the helium method. They were determined in five cases (4.9, 10.6, 2.5, 8.4, and 6.2%) giving an average of 6.5%. The results of both methods were compared with the tritium actually recovered from a uranium getter where it was collected after extraction from the alloy. The yields determined in this way ranged from 95 to 124% (Tanase, 1988). These uncertainties are definitely not due to the uncertainty of how much tritium could be extracted from the target. Less than 0.01% remained in the ^{6}Li-Al alloy targets after heating them to 700°C.

From these considerations it can be concluded that tritium control cannot rely solely on predictions of tritium production rates or indirect measurements. This applies both to breeding and inadvertent production. Rather, the controls have to make sure by containment and surveillance that no tritium can be diverted even before the first physical inventory is taken by the inspectors to determine the baseline for subsequent accountancy activities with high accuracy.

3.4.3 Measurement accuracy in inventory-taking

Inventory taking of storages for tritium gas

Bulk quantities of tritium are most easily determined in the elemental gas form. There are two options for routine measurements of bulk tritium gas inventories of storages.[30] One is calorimetry, the other is pressure–volume–temperature (PVT) combined with concentration measurement (mass spectrometry — MS, gas chromatography — GC, or beta scintillation detector — BSD).

The PVT/c inventory method[31] is based on the ideal gas law. The tritium content m_T in grams, present as T_2, is calculated by

$$m_T = M_{T_2} \frac{x}{z} \frac{p\,V}{R\,T} \tag{3.5}$$

where M_{T_2} is the molecular weight of T_2 ($M_{T_2} = 6.032$ g/mole), c is the mole fraction of tritium, z is the compressibility factor ($z = 1$ for low pressure and ambient temperature conditions), and R is the molar gas constant ($R = 8.31$ J/(mole K)). Tritium quantities present in the form as other compounds (TD, TH, T_2O, NT_3, etc.) are calculated accordingly.

In order to determine the inventory of a metal getter bed by PVT/c, the tritium has to be recovered as gas by heating the metal to temperatures well above the sorption temperature and swept into the measurement tank by means of an inert gas. After desorption of storage beds to tanks of known volume pressure, temperature and isotopic composition are determined.

This process encounters the problem that successive tritium absorption–desorption cycles change the getter material and may affect calibration.[32]

Another drawback of PVT/c is that it is very time consuming (several hours to days). For example, discharging of a uranium bed storage with a maximum capacity of 5.7 g tritium takes 2.5 hours (limited by thermal capacity); reloading takes 15 minutes (Kurz, 1984). Another source reports that complete regeneration of a uranium getter bed (with a design capacity of 1000 $L(STP)$ of hydrogen) for inventory takes about 8 hours (Ellefson and Gill, 1986).

Calorimetry is based on the decay heat of tritium. The temperature increase of a uranium getter storage is measured and compared with calibration curves, which are recorded while the storage is electrically heated. The amount of tritium can be calculated with the equation

$$m_T = \frac{P_T}{P_S} \tag{3.6}$$

where P_T is the power output of the measured item and P_S is the power output of tritium ($P_S = 0.33$ W/g). In the case of a tested prototype vessel the temperature increase was 0.4 K per 0.033 W, corresponding to 4 K/g (Tritium) (Kurz, 1984).

For samples much smaller than 1 g of tritium, microcalorimetry can be applied. By measuring the heat flow due to the decay heat of tritium introduced into a small measurement cell, the tritium content of the sample may be determined. The main difficulty with this method lies in the calibration.

A comparison of PVT/c and calorimetry was carried out by the measurement control program in the tritium enrichment area at Mound Laboratory at Miamisburg, Ohio. The results are applicable to a wide variety of gaseous operations and are compared in this section to findings of other researchers. The findings are listed in Table 3.3. It was found that PVT/MS measurements with their uncertainties completely overlapped the calorimeter measurements with their uncertainties (Lindsay *et al.*, 1987).

The relative overall accuracy of the PVT/MS technique can be calculated by using a square root of the sum of squares method:

$$\delta_I = E_1\, A = \sqrt{\delta_p^2 + \delta_V^2 + \delta_T^2 + \delta_x^2} \tag{3.7}$$

This equation is derived from equation (3.5). From the data given for the tritium facility at Mound (Lindsay *et al.*, 1987) one gets:

$$\delta_I = \sqrt{0.6\%^2 + 0.3\%^2 + 0.3\%^2 + 0.2\%^2} = 0.76\%\ (3\,\sigma)$$

The percentage uncertainties were taken at the following points which represent normally expected values: pressure at 50% of full scale, volume at full scale, temperature at midrange (298 K), and mass spectrometric composition at 90 mole%. For comparison, another study finds (Stern, 1988): $\delta_I = 0.90\%\ (2\,\sigma)$.

In Lindsay, Sprague and Brandenburg (1987), the average uncertainty in measuring the inventory of 50 bottles is reported to be ±0.91% of the total grams in the tank (i.e., about 0.15 g of tritium for 16 g bottles). Thus, the actual value of uncertainty compares well with the expected uncertainty derived by the error propagation method. The conclusion of Lindsay et al. is that "an average" tritium PVT/MS system has an accuracy of about ±1% ($3\,\sigma$) with enriched tritium (>90%).

To be on the safe side, a value of 1% is used in Table 3.3. Yet, two studies undertaken for two large European tritium-handling laboratories with inventories of up to 100 g conclude that inventory measurements based on PVT/c will have an accuracy of 3–4.5% (Kraemer, 1993; Housiadas *et al.*, 1993). This discrepancy can be understood by the fact that these two facilities are under tritium control by EURATOM. They published cautious estimates in order to make sure that the accountancy demands posed by the inspecting agency will not be too strict, which might cause problems for them. Experience shows that operating organizations fear not being able to achieve the published accountancy capabilities.

The published and above-mentioned figures are calculated using propagation of error or results which were achieved under favorable conditions. In practice, especially in routine analysis, these "theoretical" accuracies are very difficult to reproduce. The combination of different measurements conducted under specific conditions causes additional systematic errors. It should be noted that measurements which do not follow identical procedures under identical circumstances may lead to discrepancies.

The accuracies discussed so far are related to PVT/c measurements. For large inventories better accuracies are achievable with calorimetry. Its accuracy and sensitivity depend mainly on temperature measurement. The Savannah River Laboratory uses special Brown resistance thermometer bulbs with an accuracy of ±0.1 K at 298 K (Stern, 1988). The temperature measurements before and after tritium loading add up to an uncertainty of 0.2 K (0.07%) for the temperature increase ΔT. As a consequence, the uncertainty for tritium inventory would be dA/A = 0.07%. At Mound the uncertainty of calorimetry is reported to be ±0.135% of the measured value (two-tailed, 0.01 confidence) (Lindsay *et al.*, 1987), e.g., about ±0.02 g for 16 g samples. The smaller the tritium sample, the more inaccurate are the measurements. For results with smaller quantities, see Table 3.3.

Table 3.3 Expected measurement accuracies δ_E and expected accountancy capabilities E associated with closing a tritium balance.

bulk facility type	A [g]	δ_I	δ_E	E [g]	ref.
			gaseous tritium		
uranium getter storage, PVT/MS in ETHEL	0.1–10	0.045	0.06	0.02–2	(Housiadas *et al.*, 1993)
uranium getter storage, PVT/MS at TLK	1	0.03–0.04	0.04–0.06	0.13–0.20	(Kraemer, 1993)
uranium getter storage, PVT/MS at Mound	16	0.01	0.015	0.79	(Lindsay *et al.*, 1987; Stern, 1988)
uranium getter storage, calorimetry at Mound	16	0.0014	0.002	0.11	(Lindsay *et al.*, 1987)
uranium getter storage, calorimetry at ETHEL	5	0.002	0.003	0.05	(Housiadas *et al.*, 1993)
	3	0.01	0.015	0.15	(Kraemer, 1993)
	1	0.02	0.03	0.10	(Miller, 1993)
	0.15	0.05	0.07	0.035	(Housiadas *et al.*, 1993)
	0.10	0.10	0.15	0.05	(Miller, 1993)
small sample, microcalorimetry	0.014	0.003	0.004	0.0002	(Genty, 1973)
ETHEL and TLK (without waste), PVT/MS	100	0.03–0.045	0.05	16.5	(Housiadas *et al.*, 1993; Kraemer, 1993)
ETHEL and TLK (without waste), optimum mix	100	0.007	0.01	3.3	
Thermal Diffusion Tritium Enrichment Facility, Mound	100	–	0.0025	0.82	(Lindsay *et al.*, 1987)
			tritium in waste stream		
Tritium Aqueous Waste Recovery System (TAWRS), Mound	0.06	–	0.2	0.04	(Sienkiewicz, 1988)
ETHEL and TLK, waste stream	1.3	–	0.2	0.86	(Housiadas *et al.*, 1993)

Microcalorimetry allows very accurate measurement of smaller tritium quantities. It was the method by which calibration standards were originally produced (NCRP, 1976). After repeated measurements of the same substance, a tritium value of (13.56±0.04) mg at 99% confidence limit was obtained (Genty, 1973).

For large inventories the better accuracy can be achieved by calorimetry. Below a quantity of about 1 g, PVT/c is the more accurate method. Assuming the optimum mix of methods is applied at a large tritium-handling facility, the expected accuracy in closing a material balance can be estimated as 1%. It is important to note that there is a large facility (Thermal Diffusion Tritium Enrichment Facility, Mound), which achieves an even better accountancy performance ($\delta_E = 0.25\%$) (Lindsay *et al.*, 1987). In this case the expected accountancy capability E remains below the significant quantity of 1 g. In the case of a large facility (inventory of 100 g), E would be 3.3 g assuming an optimum mix of methods to measure the inventory. This would be more than one significant quantity. However, improvements in inventory determination can be expected (see following subsection) and in nuclear safeguards at large bulk handling facilities the accountancy capabilities are similar (see Section 3.4.5).

Shipper/receiver discrepancies and opportunities for improvements in accountancy

Significant shipper/receiver discrepancies appeared in 1989 with tritium supplied by the Oak Ridge National Laboratory.[33] On July 21, 1989, the U.S. Department of Energy (DOE) suspended sales of tritium because in several shipments discrepancies appeared in the amount of tritium sold to a British company and the amount it received. The total discrepancy came to 2.5 g (900 TBq). A similar problem was found at other companies and all discrepancies in the DOE's sales program added up to five grams (Broad, 1989).

DOE resumed commercial sales a few weeks later. At the end of October 1989, DOE again suspended commercial sales. In a test shipment from one building to another at Oak Ridge, workers lost 2.2 g or three-fourths of the shipment. Leakage and procedural problems were officially ruled out (Arms Control Reporter, 1989). These cases and some possible explanations for discrepancies are discussed in Section 1.4.2.

Inventory measurements can only be achieved in a reproducible manner when operations are carried out according to defined procedural guidelines and when system constants like sampling volumes and storage hold-ups are well known and periodically updated. This assures that predictions on unmeasured transfers and process hold-ups can be made. Differences in measurement procedures at the beginning and the end of the shipment are likely to result in a discrepancy, because there exist no regulations or standardizations for the measuring procedure of inventory taking (see, e.g., Broad, 1989). There are not even quantity standards or reference values available against which measured values can be compared or which can be used for instrument calibration. Without any standards it is difficult to prove that a measurement system is and remains accurate. The only existing tritium standards are those of composition. They are expensive, extremely time consuming to prepare, and available only in small quantities.

Since the amount of tritium gas trapped in the storage is of significance and since external parameters (e.g., volume of external parts such as expansion tanks, valves, tubes; temperature constance of external parts; composition of released gases) may vary, a technique for direct determination of the inventory, i.e., without removing the gas from the storage, is recommended. Calorimetry, which is based on the heat released by decay of tritium can be used for this purpose. A further advantage of calorimetry is that it eliminates the problem of sample inhomogeneity and that the isotopic composition, purity, and the chemical form in which the tritium is present need not be determined for the measurement. This method, however, is time consuming and requires large samples. With tritium quantities above 1 gram, it achieves satisfactory accuracies (see Table 3.3). To facilitate the application of calorimetry, care should be taken in the design phase of facilities that inventories of large tritium samples could be determined inside a glove box without removing the sample from the primary system.

Inventory taking in the aqueous phase

In most applications tritium is desired in the gaseous form. Since the reduction of oxidized tritium is a very complicated and expensive procedure, tritium in the aqueous phase will in most cases either be waste or tritium inadvertently produced during normal operation of nuclear reactors causing a radiological hazard. Relevant quantities of tritium in the aqueous phase occur in the coolant and/or moderator of heavy water reactors and in various aqueous waste streams from spent fuel reprocessing plants.[34] The principal sources of tritium are neutron absorption by deuterium in heavy water and ternary fission in the fuel, respectively (see Sections 2.4.4 and 2.4.5). To detect unreported diversion of tritium from these sources, the tritium concentration in heavy water reactors and the extraction of tritium at reprocessing plants have to be permanently monitored. As of now, only six large-scale industrial and research facilities which extract tritium from tritiated heavy water are in operation worldwide (see Section 2.4.4 and Appendix A), and tritium at existing reprocessing plants is always released into the environment. At future reprocessing plants, tritium extraction may become necessary for environmental reasons.

The tritium content is determined by measuring the volume and concentration of tritiated water. The most common technique for measuring the tritium concentration is liquid scintillation counting of discrete samples. At high concentrations, the minimum accountable unit for tritium in the aqueous phase can be as high as 0.01 g, because it is necessary to take a sample of a specific volume to determine the content of the tritiated water. The accuracy depends on calibration, sample preparation, and the statistical counting error. The latter dominates at the detection threshold, which lies around 10 Bq/l. For example, at a concentration of 70 Bq/l in a sample of 8 ml, an accuracy of 4.6% can be achieved when measuring for 100 min and at a blank count rate of 0.17 s^{-1} (Hessisches Ministerium für Umwelt (HMU), 1983). At tritium concentrations of 0.02 to 0.2 GBq/l, an accuracy of 20% has been achieved in aqueous waste streams (Sienkiewicz, 1988).

Accountancy of continuously running processes

In order to achieve the desired accountancy goals at tritium-handling facilities, routine operation of the system would have to be stopped at certain time intervals and the tritium would have to be transferred to a separate storage container to measure the quantity (e.g., by PVT/GC). This is referred to as discontinuous or process interrupting analytical control.

If, however, after a shutdown for measurement purposes, the start-up takes a very long time and process equilibrium is only reached within a few months, frequent process interruptions will be not be tolerable.[35] Therefore, it is desirable to develop "on-line" or "noninterrupting" methods in order not to impair the performance of operating processes by analytical control measurements.

On-line instruments can be installed, but this tends to be at the cost of accuracy. The Tritium Systems Test Assembly (TSTA) at Los Alamos can serve as an example (see, e.g., Tritium Test Facility, 1982). TSTA consists of a large interactive gas loop that simulates the proposed fuel cycle for a fusion facility (e.g., TFTR, INTOR, or ITER). Its inventory is 150 g and the gas loop is designed to handle up to 1800 g/d of DT. The various analysis instruments are located at the input of the Emergency Tritium Clean-up System, the Tritium Waste Treatment System, the Fuel Clean-up System, and the Isotope Separation System. Off-line instruments are designed to provide 0.1% analyses or better. On-line instruments offer reproducibility of at least 1% (Nickerson, 1982).

In the case of complex tritium-handling facilities, it may not be possible to assess the inventory without modelling. This can be achieved by measuring certain parameters at key positions combined with computer simulation of the material balance in components of the facility.

Different approaches lead to the development of process models for fusion reactors on a rather abstract level. They are discussed in Section 3.4.2, showing that their model predictions are too inaccurate to be of use for accountancy purposes. Thus, the tritium accountancy of specific installations will not profit much from the complete fuel cycle model. Accountancy simulation and modelling of measurement procedures which cover the details of very specific installations of a tritium laboratory is in an early stage (Gabowitsch and Spannagel, 1989). Four examples are presented here:

1. Gabowitsch and Spannagel developed the computer code KATRIA (Karlsruhe Tritium Accountancy), which allows some subsystems encountered in the Tritium Laboratory Karlsruhe (TLK) to be modeled. Special versions of the code consider the accountancy effectiveness for different accountancy frequencies or different amounts of waste (Gabowitsch and Spannagel, 1989).

2. Avenhaus applied a simulation for material accountancy to the Tritium Laboratory Karlsruhe (TLK) (Avenhaus and Spannagel, 1988). Avenhaus did not consider the possibility of diversion of tritium and thus he defines the objective of tritium accountancy as the efficient, timely detection of tritium anomalies. The simulation for tritium accountancy is of particular interest for experiments related to fusion reactor research as well as for a complete tritium fuel cycle.

3. Another process simulation program to model fusion fuel and hydrogen isotope processing systems is already commercially available. FLOSHEET, which is offered by the Canadian Fusion Fuels Technology Project (CFFTP), is a microcomputer-based program (CFFTP, 1988). The user selects process modules (e.g., cryogenic distillation, electrolysis, stream mixing) from the library, specifies how they are interconnected, and defines the relevant process unit and stream parameters. The simulator then calculates the performance of the process units to evaluate local and global mass and heat balances.

4. A concept of how to assess the tritium inventory of an operating tritiated water system (namely the Tritium Aqueous Waste Recovery System (TAWRS) at Mound) without disturbing the process was developed. It combines model calculations with the measurements of a few parameters (Sienkiewicz, 1988). A model was set up for the process in order to establish a basis for the determination of the total system inventory as a result of the two known concentrations, the tritiated feed and product streams. In addition, the tritium concentration in the various components of the system could be calculated. A goal of ±20% (control limits), expressed as the average percent difference between the actual inventory and that predicted by the model,[36] was established as the acceptable criteria for determination of the total inventory using the model. Four out of 16 comparisons of actual inventories and calculations were beyond the control limits, but further investigation showed that this could be attributed to erratic sample results. Feed stream concentrations ranged from 0.02 to 0.2 GBq/l (Sienkiewicz, 1988).

Thus, it can be concluded that even if some measured parameters are integrated in computer modelling, the accuracies have still to be improved to be of relevance for tritium control.

Tritium in waste

The most difficult task for accountancy is to assay with satisfying accuracy the quantity of tritium in disposed waste. In general, an accuracy of 20% is assumed. Vance *et al.* used a method that could determine tritium in 100 mg metal samples (Vance *et al.*, 1979). Due to the lack of suitable reference materials the accuracy of this method could only be estimated to be ±15%, assuming that tritium is homogeneously dispersed through the bulk material and that a representative sample is chosen. At the European tritium-handling facilities ETHEL and TLK, annual waste streams of some 1.3 g of tritium are expected. The goal for accuracy in determining the tritium content in these wastes is 20% (Housiadas *et al.*, 1993). This compares well with the value of 25% for the expected measurement accuracy δ_E of a separate nuclear waste storage in nuclear safeguards (see Table 3.2). As long as waste streams are small, this poor accuracy will not cause a problem for tritium controls.

However, large quantities (>100 g (37 PBq) per year[37]) of tritium in solid tritiated wastes can be expected in discarded parts from normal operation of proposed fusion reactors. Hence, the use of fusion energy, if ever realized, would cause a problem for tritium accountancy.

3.4.4 Tritium sink analysis

Within the production chain, during the storing, handling, and processing of tritium there are several sinks for this material. These sinks have to be identified and the quantity of tritium lost has to be estimated by material sink analyses.

Besides the obvious objective to minimize tritium sinks, the crucial question for the purpose of tritium controls is whether it is possible to detect and measure most of the sinks and to minimize the remaining material unaccounted for (MUF).

Not all sinks of tritium cause MUF. For example, losses due to radioactive decay can be calculated. Some of the tritium which is hidden in the equipment can be recovered. Tritium from unmeasured transfers might appear in the accountancy of another MBA. Normal loss of tritium during operation of a process, for example via ventilation systems and radioactive waste, is continuously monitored by special monitoring equipment and provides output data for tritium accountancy. Therefore, it has to be distinguished whether losses are accounted, unaccounted, or temporarily hidden. The latter two may cause problems for tritium control because they are unavailable for accountancy. These sinks have to be minimized.

Unspecified loss in the production line

The question is how much tritium may be lost (or diverted) after its generation but before the baseline of freshly produced tritium is established for accountancy.

Tanase *et al.* reported on the breeding of tritium in batches of 40 TBq ($\simeq$0.1 g) and 3 TBq ($\simeq$8 mg) in their experiments (Tanase, 1988). In an earlier experiment, about 0.5% tritium leaked through the innermost wall of the vacuum furnace and smaller amounts of tritium were released from the other components. This could be significantly reduced in the later experiments and the largest quantity escaped when about 0.01% tritium was released to the inside of the cell. This material could be completely collected with the tritium removal system (TRS). Thus, at small breeding rates, the loss can be kept close to zero.

The total loss of tritium during a large-scale production process can be estimated from the global yield g of the French military production facility at Marcoule, where g is the ratio of recovered tritium to the quantity of tritium equivalent to consumed lithium-6. This value is stated to be $g = 0.96$ (Hugony *et al.*, 1973). Assuming that no lithium is lost in the process, the fraction of tritium lost in the production chain is 4%. About 1.5 to 2% of the missing tritium is due to radioactive decay during processing. Other tritium sinks are diffusion of tritium through hot elements of the installation and exposure to air of the circuits when serviced. Considering all the sinks mentioned above establishes a balance close to 100% (Hugony *et al.*, 1973).

From this it follows that discrepancies between predicted breeding rates and actually measured tritium are mainly due to uncertainties in predicting the baseline (see Section 3.4.2) rather than to loss of tritium in the production line.

Unaccounted normal operating losses

Normal operating loss (NOL)[38] is tritium lost or no longer usable as a result of normal operation of the facility such as stack losses. This does not include stack releases due to component or human failure, which are treated in a subsection below. By definition, all tritium leaving the MBA during normal operation without being accounted for belongs to the category of unaccounted losses during handling of tritium.

There are a variety of material loss mechanisms. Most importantly, tritium can be lost, unnoticed, as residues in waste and as emissions to the environment, which leave the MBA without assessment of the tritium content.

In general, the tritium content of waste is assessed and accounted for. If this is the case and some portion of the tritium in the waste escapes the assessment it will be treated as a measurement error and is considered in the previous section (3.4.3). If tritium leaves the facility in waste without being noticed it belongs to the category of unaccounted normal operating losses. For instance, small quantities of tritium can leave a material balance area by normal capillary sampling of process gas operation.[39] If during an inventory period 50 samples of high purity tritium ($\simeq$100%) were drawn with each, implying a loss of V_s = (50 $\pm$ 25) cm^3 (STP),[40] the total unmeasured transfer would amount to (0.68$\pm$0.34) g (Ellefson and Gill, 1986). As soon as an appropriate loss mechanism is identified, it can be made sure that it will be accounted for.

Each preparation of a tritium target containing 220 TBq (0.62 g) of tritium in the Isotope Research Material Laboratory (IRML) at Oak Ridge National Laboratory (ORNL) causes a release of no more than 1.1–1.3 TBq (3.1–3.6 mg) of tritium. Since approximately 3.33 PBq (9.2 g) of tritium are needed in the sorption chamber to load the target, about 0.03% of tritium is lost during the process. During the tritium target cycle, typically half of the activity is lost by outgasing, dislodging, and sputtering.[41]

Stack loss is tritium released to the environment up the facility stack. This includes release due to permeation, maintenance line breaks, component outgasing, planned and scheduled stack releases, permeation through glove box valves, gaskets, etc. (see, e.g., Wall and Cruz, 1985).

During normal operation the tritium concentration in vented air and discharged effluents is monitored. Emitted quantities can be calculated and treated as output in the material balance.[42] Therefore, tritium lost to the environment without being measured can be expected to be much smaller than the total amount of tritium lost.

The tritium sales program at ORNL releases approximately 160 TBq (0.45 g) of tritium gas via the ventilation system each year. A total of approximately 74 PBq (207 g) of tritium is processed each year. Thus, the percentage released amounts to $\simeq$0.2% (Kobisk *et al.*, 1989). This is a typical value of containment performance. A recent review shows that a containment performance which loses not more than 1% of the inventory or even less than 0.01% per year can be achieved (compare Figure 3.2) (Kalinowski, 1993). Even if as much as 10% of tritium lost to the environment was undetected, this would not amount to more than 0.1 g/y in a large tritium-handling facility with an inventory of 100 g, i.e., L $<$ 0.1 g.

Radioactive decay

Tritium lost due to radioactive decay does not belong to the former category, because it can be calculated with a high degree of accuracy. Although a delay of a few days seems negligible compared to the tritium half-life of 12.3 years, it should be noted that this delay might be responsible for discrepancies in the order of magnitude of accountable quantities. To illustrate this, an assay of 10 g of tritium by PVT/MS is considered. A delay of 6.5 days between mass spectrometer sampling and sample analysis would result in an error of 0.01 g. The date t_0 of each inventory and transfer action has to be recorded in order to allow calculation of the tritium loss, which is due to radioactive decay. The current book value I at time t_1 can then be calculated using the equation

$$I(t_1) = I(t_0)\ exp\left(-\frac{(t_1 - t_0)\ ln\,2}{4490}\right)$$

In this formula the unit of time is days.

Hidden inventories

The hidden inventory is defined by unaccounted "hold-ups" in various process elements, most of it in uranium getter beds. They are common in nuclear materials accountancy. However, due to its high mobility and reactivity it is more difficult to locate tritium than solid nuclear materials.

Hidden inventories may be due to a variety of reasons depending on earlier treatment of the storage container (exposition, temperature, length of annealing time). The walls of the container and of pipes absorb tritium.[43] It may or may not be saturated with tritium from a previous filling. The container may contain small amounts of oxygen, which reacts quickly with tritium to give superheavy (T_2O) or tritiated (HTO) water. With no oxygen but N_2 present the formation of tritiated ammonia can be expected (Cheek and Linnenbom, 1958; and Gill *et al.*, 1986). It is well known that tritium reacts with methane to give tritiated methane (e.g., CTH_3), but CT_4 could also be produced by isotopic exchange of oil vapor and cracking of the oil[44] or by reaction of the tritium with carbon in the stainless steel. The container may contain traces of a getter metal which absorbs tritium. In most cases it can be expected that the hidden inventory increases during one inventory period, thus giving a positive contribution to MUF (see Equation 3.2).

A study carried out at Mound Laboratory (Ellefson and Gill, 1986; and Gill *et al.*, 1986) addressed two important cases for hidden inventories and quantified them for a typical tritium-handling system.[45]

1. Some tritium is irreversibly held in the storage material of heavy metal getter storages. This hidden inventory is sometimes called *hold-up* or *heel.* In the case of high-purity tritium ($\simeq$100%), the average gram hold-ups of tritium on three uranium getters with a design capacity of 1000 l (STP) of hydrogen were determined to be (1.6±0.7) g, (0.65±0.22) g, and (0.42±0.04) g, respectively. The largest quantity in a single measurement was a hold-up of 2.4 g of tritium. These quantities are real inventories but not normally accessible by the process inventory measurements. A separate accounting of hold-up is necessary to make sure it is included in the total bulk gas inventory.[46]

   Hold-up contributes to MUF when the concentration of tritium changes. This happens if the isotopic hydrogen concentration changes from one inventory to another or if the uranium getter bed is exposed to tritium for the first time.

2. Molecular sieves are another possible trap for unmeasured tritium thus causing MUF. In tritium-handling facilities they are employed in order to remove condensable gases, primarily water vapor, from tritium-containing gas streams.[47] Experiments and some calculations gave estimates of 1 to 7 g of tritium hold-ups on molecular sieves containing 1.7 kg of zeolite (dry weight) after four years of operation (Gill *et al.*, 1986).

   There are two methods to obtain tritium inventory estimates for aluminosilicate zeolites molecular sieves which can be applied without removing the molecular sieve from its working position: controlled partial regeneration and isotopic exchange.[48]

Studies for two large European facilities conclude that a few grams of tritium will be bound, which is not linearly dependent on the total inventory.[49] Most of this is available to physical inventory taking because a large fraction can be regained by special heat treatment without opening the equipment, and all can be measured by calorimetry provided the respective part of the equipment can be removed and fits into the calorimeter. Furthermore, annual changes of bound inventory are only a fraction of the total hidden inventory, i.e., $|\Delta H| < 1$ g (see Equation 3.2).

Even if the hidden inventory is not recovered for measuring, it can be estimated and preliminarily booked. For instance, the tritium control and accountability instructions at Sandia National Laboratories Livermore (SNLL) require that materials contained within the dryers remain on the accountability records at book value pending disposition. When the dryers are regenerated, any difference between material recovered and the book value is inventory difference (MUF) and is reported according to DOE procedures (Wall and Cruz, 1985).

Tritium in manufactured products

Another loss mechanism is the distribution of a significant quantity of tritium to a number of products which all contain small quantities and which are removed from being accessible to accountancy. For figures of tritium in commercial products see Section 1.3.1.

Manufacturers of commercial consumer products containing tritium typically process some 1 TBq to 100 TBq (2.8 to 280 mg) in one manufacturing lot. Thus, manufacturers may well have annual throughputs exceeding 1 g/y and have to be included in verification procedures.[50] However, accountable quantities of tritium are dispersed in many single products, each containing nonaccountable quantities of tritium. As of now there is no recycling system implemented for such products and they will end up in ordinary waste disposals or incineration plants.

The situation may be similar in research. For instance, in material research experiments, solid tritiated samples of material may be produced that contain accountable quantities of tritium. These samples may be cut up into many different sizes and shapes that eventually reduce the maximum quantity of tritium in any one piece to less than the accountable quantity.

Tritium lost in this way is not practically accessible for military purposes in significant quantities.[51] The control procedures can therefore be terminated after determining and checking out the quantity that is distributed in the products. Care has to be taken to verify the amount of tritium which ends up in the products by taking random samples.

Accidental tritium releases

Because tritium is a gas it could easily be released undetected just by opening a valve and thus causing a MUF. However, radiation protection regulations pose requirements on both significant quantities and detection time, which are far stricter than necessary for verification of nonproliferation.

Nevertheless, accidental releases do occur.[52] The question is what kind of problems do they cause for tritium accountancy. If the release passes a flow-through ionization chamber, the lost amount of tritium can be recorded. But this might not always be the case, especially if somebody tries to disguise a diversion by claiming that he or she has lost the tritium by accidental emission. Even if no recordings of effluent monitors are available, any claim can be checked at least qualitatively on the basis of environmental and bioassay sampling by inspectors and by modelling the atmospheric dispersion after the claimed release.[53]

3.4.5 Conclusions on the efficiency of tritium accountancy

A number of weak points in tritium accountancy have been identified. To ensure adequate verification additional measures have to be taken.

1. The determination of the quantity of freshly produced tritium by calculation bears a considerable uncertainty. Therefore, the nonremoval of tritium right after its production and before the initial physical inventory has been taken to determine the baseline for accountancy has to be ensured by adequate measures of containment and surveillance.

2. Accidental emissions without recording can be verified only qualitatively. Therefore, it has to be ensured that tritium is safely contained, significant quantities are distributed on multiple storages, and emissions are monitored adequately. These measures contribute to radiation protection as well. If need be, the concentration of tritium in the environment can serve as a rough check of tritium spills.

3. The accountancy of large amounts of tritium in waste (5 g/y and more) might be a problem. Therefore, the amount of tritium in waste has to be minimized. This is welcomed for economic reasons as well but may be impossible to achieve if fusion energy were realized. In that case, containment and surveillance of waste become the only reliable method of choice.

An extensive analysis of the accuracy in tritium inventory measurements in different modes (gaseous and aqueous storages, operating systems) was undertaken to estimate typical values of MUF when closing a material balance.

It is interesting to note what requirements on accuracy have been proven to be realistic and achievable in practice. Such values can help to evaluate the accountancy capability and to decide whether accountancy would be an adequate approach for tritium control.

Most measurement accuracies of MUF which are actually achieved or expected in various tritium-handling facilities are between 0.0025 and 0.05 and about 0.2 for tritiated waste.[54] Studies undertaken for two large European tritium-handling facilities with inventories of up to 100 g conclude that inventory measurements will have an accuracy of 0.03 to 0.04% (based on PVT-c), and with current technical constraints the MUF will have a standard deviation of about 5% (Kraemer *et al.*, 1993; and Housiadas *et al.*, 1993). But in the previous sections these assessments are shown to be over-conservative, and major improvements in accountancy can be expected, especially through the application of calorimetry and by a more accurate determination of tritium in solid waste.

The measurement control program of the Tritium Enrichment Facility in the Mound Laboratory that can be used as a reference achieves high accuracy: a difference of just 0.25% between the measured inventory and the sum of all incoming transfers minus the sum of all outgoing transfers would be cause for investigation (Lindsay *et al.*, 1987).

The contribution of waste to the unaccounted tritium can be expected to be small because the total waste per year is typically in the order of 1% of the inventory. Even for a large facility the expected accountancy capability E associated with closing the tritium balance in the waste stream will be less than 0.1 g/y.

Even if as much as 10% of tritium lost to the environment was undetected, this would not amount to more than 0.1 g/y in a large tritium-handling facility with an inventory of 100 g, i.e., $L < 0.1$ g.

Most of the hidden inventory can be regained by special heat treatment without opening the equipment, and annual changes of bound inventory are only a fraction of it, i.e., $|\Delta H| < 1$ g. Table 3.1 compares the expected accountancy capabilities E for plutonium and tritium.

The expected accountancy capability for the imaginary closing of the material balance for the total world inventory of separated civilian plutonium is between 370 and 740, significant quantities depending on the achievable measurement accuracy (see Table 3.1). This means that diversion of plutonium for hundreds of nuclear weapons during one inventory period may remain undiscovered. In nuclear safeguards, according to the NPT or the IAEA statute, the accountancy verification goal is related to but not necessarily equal to one significant quantity. At smaller facilities the error limit may be a fraction of a significant quantity, but at a large facility it may be the equivalent of several times the material needed to produce a weapon. This can be sharply criticized, but as a matter of fact this is still accepted officially to guarantee "adequate" verification.

The expected detection capability in an imaginary closing of the material balance of the whole world inventory of all separated, civilian tritium stocks would amount to some 66 to 1300 SQs of tritium.

When considering that the SQ for tritium is defined here much more conservatively (at least by a factor of 5) compared to the official SQ of plutonium, it can be concluded that with current state-of-the-art technology and under normal operating conditions, tritium accountancy is possible on a routine basis with a capability which compares well with that required for nuclear safeguards.

As a result, control tasks of Type II (verification of nonremoval) are feasible and can be based on technologies and experience from tritium accountancy already established for radiation protection purposes. The precedent for this type is given by an agreement between EURATOM and Canada.

3.5 Containment, surveillance, and physical protection

3.5.1 Containment and surveillance

Containment and surveillance are control procedures which are implemented to supplement and strengthen tritium accountancy (the main measure to verify non-removal) as well as verification of nonproduction. Typical instruments for containment and surveillance are video cameras, seals, and tags. These methods are applied in nuclear safeguards to detect activities related to plutonium breeding (e.g., insertion of breeding targets made of natural uranium) as well as to removing fissionable materials (e.g., removing part of the fuel). Most of those methods which are already in place for nuclear safeguards would be useful for tritium controls as well.

Containment and surveillance applied to fresh fuel and fuel in the reactor core is a method to verify that no targets for tritium breeding are inserted or removed.

Containment and surveillance applied to spent fuel is suitable to verify the nonremoval of tritium in fuel originating from ternary fission until the fuel is reprocessed or until safeguards are terminated due to final disposal which is not yet available.

If containment and surveillance of heavy water is implemented[55] and the tritium concentration is additionally monitored, the nondiversion of tritium contained in this water can be verified.

Tritium storage and shipment containers can be sealed in the presence of an inspector after they have been filled with a verified quantity of tritium.

For more examples of containment and surveillance methods in tritium control see Section 3.7.

3.5.2 Physical protection

Apart from tritium containment and surveillance, physical protection is an additional approach to building barriers to diversion of tritium. Physical protection comprises "technical barriers" associated with the physical and chemical constitution of tritium as it occurs in the equipment of tritium facilities and "physical security barriers" such as alarming and automated monitoring systems, seals on containers, security guards, and personnel screening. Even in the case of proposed fusion power reactors, physical protection seems to be feasible. Burning, breeding, and recycling will all be integrated in the reactor plant, eliminating the necessity of tritium shipments except for the initial inventory needed to start up new reactors.

A typical task of physical protection is to detect tritium in shipping containers at check-out points of facilities. When physical protection of nuclear materials is implemented, portal monitors based on neutron counting can detect the passing of small quantities of fissionable material (down to a few grams of plutonium).

In contrast, it is comparatively easy to carry or store tritium without being noticed, because it is difficult to identify that a vessel contains tritium.

Due to the low energy of the β^--electrons, the detection of tritium in air and in liquids is particularly difficult compared to most other radioactive isotopes. This is different for tritium in metals. The induced soft X-rays (bremsstrahlung) can be detected. The continuous X-ray spectrum is superposed by peaks at characteristic energies. X-rays emitted from metals are not proof of the presence of tritium, because they can be induced by several other processes. If, however, X-rays with energies below 18.6 keV can be detected and there is a rapid decrease above this energy, this would be an indicator of the presence of tritium, because the maximum β-energy of 18.6 keV is a specific constant of tritium. But the penetration depth of the induced X-rays is in the order of 10 μm. After penetrating an aluminium foil of 15 μm thickness, the intensity is reduced roughly by a factor of 2. Therefore, any radiation signals can easily be shielded by the diverter.

However, it is not possible to develop containers that do not release tritium, because gaseous tritium permeates, at varying rates, through normal materials of construction. IAEA regulations limit releases of tritium from transport containers to 10^{-5} Pa l/sec. Ordinary steel barrels have leak rates of 10^{-1} Pa l/s and $<10^{-4}$Pa l/s after being welded tightly (Luthardt and Huschka, 1983). Uranium getter storages achieve lower leak rates, but tritium emissions may still be detectable provided that air samples can be taken from an unvented room in which the container has been placed for a sufficient length of time.

Active interrogation could also be used to identify tritium in storages from the outside, especially if a tritium storage is declared to be empty but cannot be opened to analyze the content. Neutron radiography has been proven to be a suitable method for the nondestructive analysis of metals for tritium. In fact, not tritium but its decay product helium-3 has been detected. Helium-3 has a very high absorption cross-section for thermal neutron (5333 barn) in comparison to metals used typically for gettering tritium (Buchberger *et al.*, 1989). But to allow interpretation of radiographs, a reference picture of the empty storage may be required.

3.6 Detection of clandestine facilities and activities

A major inefficiency in current nuclear safeguards lies in the very low probability of detecting clandestine facilities which are operated for nuclear weapons purposes. Current IAEA-type safeguards are basically cooperative verifications on a routine basis. An additional approach has been proposed which makes use of international technical means (ITM) to detect indications of clandestine facilities or activities. To some degree, existing national technical means (NTM) for space and airborne remote sensing are capable of detecting clandestine facilities (see, e.g., von Hippel and Levi, 1986; and Jasani and Stein, 2002). Tremendous efforts have been put into the development of remote sensing capabilities for military purposes, especially for reconnaissance and smart sensors on "brilliant pebbles" for SDI. New technologies, especially in microelectronics and data processing, promise a further major step in this development, which can be exploited for verification purposes. The detection of suspicious sites could then trigger further on-site inspections to scrutinize the alleged illegal activities. The IAEA has the right to conduct special inspections according to §73 and 77 of INFCIRC/153, the verification agreement required in connection with the NPT.

In fact, in the search for and investigation of clandestine nuclear facilities, remote sensing has already played a crucial role. It appears, however, that only photographic picture evaluation, infrared imaging from satellites as well as aircraft, and perhaps air sampling have been used. Two examples for the identification of nuclear facilities are Algeria's Oussera complex and the UN missions to Iraq. Satellites have played a crucial role in bringing the Oussera complex to light and revealed considerable evidence that it has military significance and may be nearly completed (Gupta, 1992). The UN inspections in Iraq relied heavily on the interpretation of photographs taken from an American military satellite and from an American U2 aircraft which undertook several flights from Saudi Arabia.

Both examples tell the same story. Currently available intelligence capacities including military satellites failed to detect and interpret the militarily motivated nuclear activities: Algeria's Oussera complex escaped attention for eight years, and Iraq's huge weapons program was significantly underestimated. Only with highly intrusive inspection activities imposed on a loser of a war by UN resolution 687 was it possible to undertake the necessary flights. This would not currently be acceptable under the NPT.

Nevertheless, the options for aerial reconnaissance to verify arms limitation agreements have gained international attention and acceptance (see, e.g., Banner *et al.*, 1990). Therefore, it is worthwhile to study the opportunities they provide for NPT safeguards and international tritium controls.

The specific strength of satellite-based sensors is their capability to locate and characterize facilities, but they do not provide indications of the nuclear isotopes which are handled at the facility in question. In particular, it is not possible to find out whether tritium is present at the location. Infrared satellite images can be used to detect heat sources. Clandestine production reactors and possibly lithium enrichment plants might be detected by interpreting these photos.

The specific strength of the analysis of remote environmental samples is to provide information on the kind of activities (e.g., those involving tritium) and relate them to locations by following air trajectories or water streams backwards. Remote monitoring of environmental radioactivity provides data and off-site sample analyzing methods provide opportunities which could be used to gather proliferation indicators. For the special inspections in Iraq the IAEA used chemical analysis of environmental samples to find proliferation indicators. Under the Additional Protocol the IAEA applies environmental sampling within or close to a facility. The application of Wide Area Environmental Monitoring is under investigation.

Since tritium is a very mobile gas it is not possible to contain it completely. Tritium production and handling involves several processes during which a loss or leakage of tritium cannot be avoided. Hence tritium emissions can be used as an indication of clandestine tritium production and handling facilities. Emissions from facilities handling large quantities of tritium (e.g., facilities at which tritium is extracted from target slugs) can even be used as semi-quantitative measures of the amount of tritium handled, because the emissions originating from this process are very large and roughly proportional to the total amount treated. This relationship becomes evident from Figure 3.2, a diagram which shows tritium emissions vs.

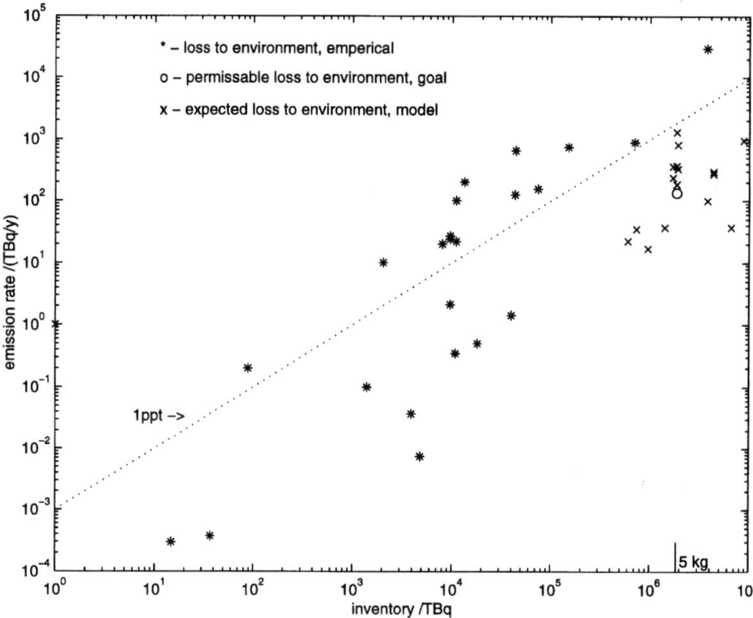

Figure 3.2 Containment performance of different tritium systems under normal operational conditions.

tritium inventory.[56] This relation was used by Cochran *et al.* to estimate from published emission values the amount of tritium production at the Savannah River Plant in South Carolina, U.S. (Cochran *et al.*, 1987). At facilities where smaller amounts of tritium are handled, the emissions to the environment are not as large, but they can still serve as an indication of tritium handling, especially by comparing it to the krypton-85 level.

The tritium/krypton-85 ratio can indicate the type of the emitting facility. It is highest for emissions from tritium production or handling facilities, and lowest for reprocessing facilities. However, large variations between the emissions of different facilities requires detailed data for the particular situation, and appropriate situations have to be selected where measurements of tritium/krypton-85 ratios may be useful. On the other hand, the tritium/krypton-85 ratio in liquid effluents from nuclear facilities is at least three orders of magnitude lower than in undisturbed continental or ocean surface waters (Schröder and Roether, 1979).

Tritium is absorbed by the soil and vegetation, which acts as a cumulative memory of previous tritium emissions. Depending on the weather conditions at the time of emission, tritium can be found tens to hundreds of kilometers away from the emitting facility in concentrations well above the "natural" background, which still is dominated by tritium originating from atmospheric tests of nuclear weapons (see Section 1.2).

From radiation protection there is broad experience in determining tritium concentrations in aqueous effluents and gaseous emissions (see, e.g., National Council on Radiation Protection and Measurements [NCRP], 1976). Effluent and environmental monitors are commercially available or under development (Nickerson, 1982; and Wood *et al.*, 1993). The question is whether the tritium emissions of appropriate facilities would give a significant signal against the background in order to enable their detection by tritium measurements.

The concentration at the point of emission is known when the total volume of exhaust air or aqueous effluents is given. Typical exhaust volumes of large facilities (fission reactors) are in the order of 10^9 m^3/y with a water content of 10 g_{H_2O} m^{-3}. With a yearly emission of 1 TBq as tritiated water (HTO), the average concentration in air would be 1000 Bq/m^3 and the average tritium content in water vapor would be 10^5 Bq/kg_{H_2O}.

With these data a dispersion calculation has to be done in order to estimate the concentrations at a certain distance from the emitting facility. The average natural background of tritium in world waters is of the order of 0.1 Bq/kg_{H_2O}; the average tritium concentration in drinking water near CANDU stations is about 0.03 to 0.3 Bq/kg_{H_2O} (Nickerson, 1982). Detection limits are around 10^{-3} Bq/kg_{H_2O} for HTO in water and less than 10^{-2} Bq/m^3 for total tritium in air (Wood *et al.*, 1993). Without dispersion and with the data used above, the detection limits with respect to emissions expressed in activity per year are 10^{-8} TBq/y for HTO in water vapor and 10^{-5} TBq/y for total tritium in air. Assuming a containment performance of 1 ppt, tritium facilities with inventories larger than 0.01 TBq could be detected from total tritium emissions measured at the stack. This is more than four orders of magnitude less than the significant quantity.

By following air trajectories, plumes can be traced back to suspected sources at distances of several hundred kilometers.[57] It has been shown that evidence of pulsed discharges of krypton-85 and tritium from nuclear energy installations can be derived from precipitation studies at distances of several hundred kilometers (see, e.g., Weiss, 1986).

3.7 Control activities at different facility types

3.7.1 Overview of relevant facilities worldwide

Table 3.4 gives an overview of all facilities worldwide which are of relevance for international tritium control as of 1992. Not much has changed since then. All considerations regarding tritium control that are based on these historic numbers are still fully valid. The numbering of facility types is the same as defined in the diversion path analysis (Chapter 2; see especially Figures 2.2 and 2.3, Section 2.6; Appendix A).

Since tritium can be produced only in nuclear reactors or in other strong neutron sources, all facilities which are relevant for tritium production are already covered by the present nuclear safeguards performed by the IAEA. There are some facilities in which no nuclear materials are handled except tritium. The diversion possibilities of extracting inadvertently produced tritium or removing tritium from existing stocks would make it necessary to place a number of facilities worldwide under safeguards in addition to those which are already inspected to verify fissile material inventories. The additional number of tritium-handling facilities (facility types 5 to 9) probably lies between 3 and 50, depending on the specific agreement.[58]

Not all facilities which are under nuclear safeguards have to be inspected to verify an international tritium control agreement because it would not be possible to divert more than one significant quantity within the desired detection time at some of these facilities. In this study 1 g/y is taken as the decision criterion (see Section 3.2.2). This quantity may be exceeded by the maximum inventory during one year, by the annual throughput, or by the production capacity of a facility. Column two of Table 3.4 gives the total numbers of facilities which currently exist worldwide; column three gives the number of facilities which are under nuclear safeguards. Column four gives the number of facilities which would have to be inspected for tritium in addition to those of column three, in the case that tritium were added to the materials under control according to the NPT. Only facilities of states party to the NPT are counted, but not those which are in nuclear weapon states as defined by the NPT. This shows that only an additional 12 facilities are of relevance for tritium control.

The numbers given in column five assume a combination of an International Tritium Control System (ITCS) and an Integrated Cutoff Agreement (ICO) (see Sections 1.7.2 and 1.8.3). This would ban worldwide the production of any fissionable as well as fusionable materials. It is chosen as a reference for the case in which the maximum number of facilities would fall under tritium control. This number is smaller compared to the total number of existing facilities because some facilities lie below the critical threshold inventory, throughput, or production capacity. This number includes known military production reactors which are shut down (some 28; see Table 2.6) but not yet completely decommissioned and not dismantled, because they would still have to be monitored.

3.7.2 Control activities

In this monograph a model (generic) control approach is chosen (IAEA, 1987a). All control activities described in the following list are derived from postulated reference facilities from the diversion and production path analysis and the control criteria outlined above. This is done for all facility types listed in Table 3.4. Since individual facilities might have specifications which are different from the reference facility, the actual controls might differ from those outlined in this section.

Three steps are necessary to perform control procedures:

Initial activities: Negotiations on the facility attachment; verification of facility parameters and of the baseline inventory; establishing a data file and evaluation program for the facility; providing equipment needed for inspections; and installation of containment and surveillance equipment.

Evaluating reports: The facility operator is obliged to provide regular reports on the operation, on the input and output of materials, and special reports on anomalous incidents. These reports have to be evaluated and inspection activities prepared according to the results.

Inspections: During inspections, measures are taken to ensure that none of the acquisition paths has been taken without being reported. The main tasks

Table 3.4 Survey of facilities relevant for tritium control (worldwide).

	facility type [a]	total number in 1992 [b]	Pu/HEU safeguards applied in 1992 [c]	minimum added for ITCS [d]	maximum for ITCS and integrated cutoff [e]
1a)	Nuclear power reactors	424	201	0	424
1b)	Heavy water power reactors (included in 1a)	(32)	(28)	(0)	(32)
1c)	Research reactors and critical assemblies	323	169	0	267 [f]
1d)	Heavy water res. reactors (included in 1c)	(21)	(13)	(0)	(11) [g]
1e)	Military production reactors (incl. those shut down)	51	0	0	51
1f)	Special neutron sources [h]	4	1	0	4
2)	Fuel fabrication plants	42	23	0	42
3)	Separate spent fuel storage facilities	28	19	0	28
4)	Reprocessing facilities	22	6	0	10 [i]
5)	Nuclear waste storages and disposals	23	1	3	7
6)	Detritiation facilities for tritiated water	6	0	3	6
7)	Research facilities with inventory or annual throughput > 1 gram	12	0	4	12
8)	Tritium manufacturers with inventory or annual throughput > 1 gram	21	0	4	21
	total	956	420	14	872

[a] The numbers of facility types are the same as used in Section 2.6, Appendix A, and the following subsection on control activities.

[b] All figures are given as of December 31, 1992. Some include suspected facilities; some may fall short of the actual numbers because information about the existence of facilities may not be available to the authors. Not included are facilities which are planned, under construction, shut down, or held on standby except for facility type 1e). Main source: Varley *et al.*, 1993.

[c] All figures are given as of December 31, 1992. Source: IAEA (1993b).

[d] These numbers show the difference between those to be controlled under current IAEA nuclear safeguards activities and those to be controlled if tritium is added to the materials controlled under the NPT, i.e., the integration of an International Tritium Control System (ITCS) into the NPT without controlling the facilities in nuclear weapons states.

[e] These numbers assume a combined implementation of an integrated cutoff agreement

and an ITCS with all relevant states joining the respective treaties. This represents the reference case in which the maximum number of facilities would fall under tritium control including former military production plants which are assumed to be held on standby. This number is smaller than the total number of existing facilities because some facilities lie below the critical threshold inventory, throughput, or production capacity.

[f] It will only be necessary to control research reactors which can be used to produce more than a significant quantity of tritium within a year. 1 g/y is chosen as an example. In view of the production rates of dedicated facilities as given in Table 2.7, reactors with < 200 kW_{th} can probably be excluded from controls.

[g] Only research reactors with > 12 MW_{th} have to be controlled to cover the heavy water path with production capacities larger than 1 g/y (see Table 2.7).

[h] Presently, this category covers spallation neutron sources. In the future, fusion research facilities might have neutron fluxes high enough to produce a significant quantity of tritium within one year. They will then fall under this category.

[i] Only reprocessing plants with a capacity exceeding 10 tons heavy metal per year are included. Smaller facilities would probably not release more than 1 g tritium per year and might therefore not fall under tritium control (see Section 2.4).

---

are independent verification of data reported by the operator, evaluation and maintenance of containment/surveillance measures, and inventory verification. In the subsequent list, the inspection activities are described in detail.

The actual extent of control activities depends on the relevant control agreement. Only facilities in those countries which are parties to the control agreement and only those facility types covered by the agreement have to be considered for control. Parameters varying with the control agreement are the significant quantity SQ and the required detection time t_d. In the following list of inspection activities SQ is assumed to be one gram and t_d to be one year.

List of inspection activities by facility type

Activities can be on-site or in headquarters. Some of the activities described or similar activities with the same purpose are already part of routine IAEA inspections. They are marked with (∗). Some of the activities described are already performed on the facility level for radiation protection purposes. These are marked with (#). Some of the new activities will provide additional information useful for current nuclear safeguards purposes. Some of the activities suggested here may be relevant only under certain circumstances and some may add a bit of redundancy.

1a) 1b) Nuclear power reactors:

1. (∗) Item counting of fresh fuel assemblies.
2. (∗) Seal storage at fuel fabrication facility.
3. Control rods, burnable poison rods, and any other parts determined for insertion into the reactor core which are not received from controlled plants may be verified by active nondestructive analysis for the absence of lithium-6.
4. Item identification of replaced control rods. Compare the length of time they have been used with the normal lifetime.

5. (∗) Visual inspection of reactor core and fuel assemblies during reactor refuelling periods. Fuel/lithium assemblies specially designed for tritium production can be identified because they look quite different from conventional fuel assemblies. Any targets inserted in addition to conventional assemblies inside or outside the core can be easily detected by visual inspection.
6. (∗) Seal reactor vessel.
7. (∗) Seal special instruments used to exchange fuel rods.
8. (∗) Item counting and identification of spent fuel assemblies.
9. (∗) Detect dummies or vacancies from removed targets in spent fuel assembly storage pond, e.g., with Cherenkov Viewing Devices.
10. (∗) Verify spent fuel assemblies against operator-declared values of burn-up, and verify the consistency of declared cooling times for spent fuel assemblies.
11. (∗) Camera surveillance at storage pool. Filming is triggered by crane movements.
12. Take sample of coolant and moderator water and analyze it for helium-3 content.
13. At HTRs only: Monitor helium-3 content of the coolant. The normal fraction is 0.2 ppm. The tolerated limit will be about 0.4 ppm divided by the power in GW_e.

1c) 1d) Nuclear research reactors:

Everything is identical to activities at power reactors. Differences appear only in the case that tritium production is permitted, e.g., for research purposes. Only research reactors with thermal power exceeding 200 kW_{th} fall under controls. The following activities are necessary in addition to those at power reactors:

1. (∗) Seal unused target positions.
2. (∗) Verify integrity of seals on target positions.
3. (∗) Review film for target changes; compare with records.
4. In the case of additional irradiation targets: Verify the absence of lithium-6 by active nondestructive analysis.
5. In the case of rapid power excursion experiments: special inspection of the tritium produced inadvertently from helium-3, depending on the production capacity.

1b), 1d) Nuclear reactors with heavy water:

The basis for controlling the tritium contained in heavy water is provided by specific measures which are performed in addition to the planned IAEA safeguards for heavy water. The evaluation of the material balance is based on records auditing, comparison of records with reports, and the standard verification methods as applied to flow and inventory verification. Heavy water inventories can be found in the moderator system, the primary heat transport system, the upgrading system, the downgraded heavy water stocks, and in the reactor grade heavy water store (Morsy *et al.*, 1987).

1. (*) The heavy water inventory in the reactor is determined by measuring the quantities in all the reactor vessels, e.g., by using a filling level measuring device. The inventory of heavy water in containers outside the reactor is determined from measurements on randomly selected containers (Morsy *et al.*, 1987).
2. (*) Maintain containment and surveillance of heavy water.
3. (*) Measure inputs and outputs with flow meters at all connections to the heavy water container.
4. (*) Authenticate in-plant flow meters.
5. (*) Compare flow meter readout with reports.
6. (#) Continuously monitor tritium concentration in the different locations of heavy water in the facility.
7. Verify integrity of monitors for tritium concentration.
8. Compare monitor printout to reports.
9. Check consistency of concentration changes, inputs, and outputs with operational records.
10. In the case of accidental releases, conduct special inspections to check stated losses by appropriate indicators.

1e) Military production reactors:

1. Case of tritium production not being permitted:
   (a) National Technical Means to verify the status of standby reactors.
   (b) Reactor power monitoring of dual-purpose reactors running for power production. Check with electricity generation.
   (c) Further inspections may be the same as in power reactors.
2. Case of permitted production of tritium with lithium targets below a certain production limit (compare Stern, 1988). No (*) is noted here because this scenario would most probably apply to military facilities which are not yet under nuclear safeguards.
   (a) Inspector surveillance during fuel and target reload.
   (b) Continuous optical camera surveillance during reactor operation.
   (c) Seal application at fuel and target fabrication facility. The tamper proof seals are furnished with serial numbers to enable item identification.
   (d) Item identification and counting at the reactor site.
   (e) Application of seals to ensure that only measured target assemblies are placed in core.

(f) Measuring the lithium-6 content before and after irradiation to determine the total amount of tritium produced.
(g) Physical inventory verification (PIV) of tritium in the extraction tank would complement lithium-6 accountancy by providing a second check on the accuracy of lithium-6 measurements.
(h) Sample tritiated waste and verify tritium content.
(i) Audit and evaluate records and reports on tritium accountancy provided by the operator.
(j) Tritium accountancy at target processing plants.
(k) Identify, review, resolve, and evaluate loss mechanisms, shipper/ receiver differences, and MUF, as appropriate.
(l) Independent measurements to assess the quality of the operator's measurements.
(m) Check calibration of in-plant equipment.

1f) Special neutron sources:

1. Verify the absence of lithium-6 by nondestructive analyses of irradiation targets.
2. (*) Verify integrity of seals on target positions.
3. (*) Review film for target changes; compare with records.

2) Fuel fabrication plants:

Control measures applied to fabrication charges and fuel pellets are not of interest for lithium-6 detection. Only complete fuel rods and fuel assemblies are subject to tritium control.

1. (*) Nondestructive determination of total uranium content and enrichment of statistically selected fuel rods.
2. (*) All fuel rods are counted and their identification number noted.
3. Control rods may be checked for lithium-6 by active nondestructive analysis before insertion into the fuel assembly.
4. Control rods are tagged immediately after their examination.
5. The assembling of the first fuel assembly of each production series may be supervised by the inspector and its weight determined as a standard. All fuel rods of this assembly are individually scanned for the detection of non-fissile material such as lithium-6 or dummy materials that would later be replaced by lithium-6.
6. (*) The neutron multiplication rate of the first fuel assembly of each production series may be determined in order to establish a standard.
7. (*) Geometrical properties of all fuel assemblies are verified by visual inspection. This includes counting of fuel rods and empty positions as well as measuring the dimensions of the assembly.

8. (∗) The total content of fissile material is checked for each fuel assembly by use of the neutron coincidence collar. If fuel is replaced by lithium-6 or empty tubes are filled with lithium-6, a decrease of the neutron multiplication rate from fission can be detected.
9. All fuel assemblies may be weighed.
10. Identification tags of control rods inserted into the fuel assembly may be checked on a statistical basis.
11. In case of anomalies which cannot be resolved by consultations with representatives of the fuel fabrication plant, the absence of lithium-6 has to be verified. An adequate destructive or nondestructive method has to be chosen according to the specific anomaly.
12. (∗) Seal fuel assemblies in such a way that no rods can be exchanged and no targets can be inserted in transit.
13. Burnable poison rod assemblies and thimble plug assemblies may be checked by active nondestructive analysis to detect lithium-6.
14. (∗) Seal storage at fuel fabrication facility.
15. (∗) Supervise any transfers from and to the storage (e.g., filling of transportation cask).
16. (∗) Seal transportation casks.
17. (∗) Check consistency of fabrication reports provided by the facility operator.

3) Separate spent fuel storage facilities:

1. (∗) Item counting and identification of spent fuel assemblies.
2. (∗) Detect dummies or vacancies from removed targets in spent fuel assembly storage pond, e.g., with Cherenkov Viewing Devices.
3. (∗) Camera surveillance at storage pool. Filming is triggered by crane movements.
4. Visual inspection of spent fuel. If it was heated to some 1000°C to extract tritium, traces of such treatment should be detectable by visual inspection.

4) Reprocessing facilities:

1. (∗) Camera surveillance at storage pool. Filming is triggered by crane movements.
2. (∗) Item counting and identification of spent fuel assemblies.
3. (∗) Detect dummies or vacancies from removed targets in spent fuel assembly.
4. The integrity of fuel is investigated by visual inspection just before reprocessing is started. In the case of traces that could indicate any preceding heat treatment, further investigations are triggered.
5. (∗) Compare facility records to reports.

6. (#) Monitor gaseous tritium emissions.
7. (#) Monitor aqueous tritium emissions.
8. Verify integrity of monitors for gaseous tritium emissions.
9. Verify integrity of monitors of tritium concentration in aqueous waste streams.
10. Compare monitor printouts to reports.
11. Check tritium concentration in selected waste containers.
12. Check consistency of concentration changes, inputs, and outputs with operational records.
13. Containment and surveillance of aqueous high-level waste, organic and cladding waste.

5) Nuclear waste disposals:

For radiation protection purposes, the accountancy of waste packages is quite common.[59] Nuclear waste disposals for spent fuel or high-level radioactive wastes from reprocessing do not currently exist (see Section 2.6). Most existing final disposals are of no or low relevance for tritium controls.

1. (*) Item counting and identification of spent fuel assemblies.
2. (*) Detect dummies or vacancies from removed targets in spent fuel assemblies.
3. (#) Seal containers with tritiated waste.
4. (#) Item counting and seal inspection of containers.
5. (#) Monitor gaseous tritium emissions.
6. Verify integrity of monitors for gaseous tritium emissions.
7. Compare monitor printouts to reports.
8. (*) Supervise the irretrievable disposal of waste packages in which large quantities of tritium are contained and for which nuclear safeguards have been terminated.

6) Extraction facilities for tritiated heavy water:

There is no need for inside inspections at the detritiation plant. Controlling emissions and residual concentrations in the detritiated heavy water and checking the results against tritium output measurements should be sufficient. The losses and inventory changes would be determined on the basis of process-monitoring data: input feed, output streams, stack and aqueous effluent emissions. There will be one physical inventory verification (PIV) during an inspection period with length t_i and several interim inspections to verify the other components of the material balance (i.e., shipments, receipts, shipper/receiver differences and losses).

1. (#) Verify tritium concentration in the tritiated water.
2. (*) Verify quantity of tritiated heavy water in storage tanks.
3. Verify integrity of concentration monitors.
4. (#) Monitor tritium concentration and purity in facility output.
5. Verify integrity of output monitors.
6. (#) Compare monitor printout to reports.
7. (*) Heavy water accountancy.
8. (*) Containment and surveillance of heavy water.
9. Authenticate in-plant flowmeter.
10. Audit and evaluate records and reports on tritium accountancy provided by the operator.
11. (#) Verify physical inventory (PIV).
12. (#) Identify, review, resolve, and evaluate loss mechanisms, shipper/receiver differences, and MUF, as appropriate.
13. (#) Check calibration of in-plant equipment.
14. (#) Compare internal consistency of accounting vs. operational records.
15. Apply seals to tritium storages.
16. Verify integrity of seals on tritium storages.
17. (#) Check tritium concentration and purity in selected storage containers.
18. (#) Monitor tritium emissions.
19. Verify integrity of monitors for gaseous tritium emissions.
20. Verify integrity of monitors for tritium concentration in aqueous waste streams.
21. Compare monitor printouts with reports.

7) Research facilities with inventory or annual throughput > 1 g/y:[60]

1. Shipments of > 0.1 g are allowed to controlled facilities only.
2. Record and check of data received from facilities handling more than 0.1 g tritium.[61]
3. Verify integrity of monitors for tritium concentration and purity at key measurement points.
4. (#) Compare monitor printouts with reports.
5. Audit and evaluate records and reports on tritium accountancy provided by the operator.
6. (#) Physical inventory verification (PIV).
7. (#) Identify, review, resolve, and evaluate loss mechanisms, shipper/receiver differences, and MUF, as appropriate.

8. (#) Check calibration of in-plant equipment.
9. (#) Compare total book inventory with facility inventory.
10. (#) Compare internal consistency of accounting with operational records.
11. (#) Monitor emissions of tritium.
12. Verify integrity of monitors for gaseous tritium emissions.
13. Verify integrity of monitors for tritium concentration in aqueous waste streams.
14. (#) Compare monitor printouts with reports.

8) Tritium manufacturers with inventory or annual throughput > 1 g/y:

1. The quantity of any received or sent shipments of tritium above 1 g is measured by an inspector.
2. (#) Compare shipper/receiver records to reports.
3. Inventory statements are required from manufacturers and verified once a year.
4. Shipment containers are sealed after the inventory measurement, and the seal is checked before verification of the received quantity.
5. Manufacturers are required to provide an annual report indicating the type and number of products manufactured during the year and the quantity of tritium contained. These data are statistically verified. The tritium concentration and purity are checked in selected manufactured products.
6. (#) Manufacturers retain records of any sale of their products. These records are checked for consistency with production and inventory data.
7. Products containing more than a certain amount of tritium (e.g., > 10 mg) may be charged with a deposit and will be taken back by the manufacturer (or a national authority) after the end of their useful life or when they become redundant.[62] A label which explains the return duty and carries an identification number is affixed to each product in question. The manufacturer has to ensure proper recycling or disposal of the tritium contained in returned products, and has to report on these activities. Inspectors check the records and verify statistically selected items.
8. (#) Monitor tritium concentration and purity at key measurement points.
9. Verify integrity of these monitors.
10. (#) Compare monitor printouts with reports.
11. (#) Compare manufacturing records with reports.
12. (#) Compare total book inventory with facility inventory.
13. (#) Compare internal consistency of accounting with operational records.
14. (#) Monitor tritium emissions.
15. Verify integrity of monitors for gaseous tritium emissions.

16. Verify integrity of monitors for tritium concentration in aqueous waste streams.

17. (#) Compare monitor printouts with reports.

3.8 Conclusions on verification

All 55 diversion paths described in Chapter 2 are listed in Table 3.5. The numbering of the diversion paths is the same as used in Chapter 2 and in the lists above. With 50 diversion paths, a diversion rate of $SQ/t_d = 1$ g/y can be exceeded. All these paths can be covered by appropriate control activities. This can be seen from the survey of acquisition paths, concerned facilities, and appropriate inspection activities given in Table 3.5.

At nuclear reactors (facility type 1) most tritium control activities of **type I (verification of nonproduction)** are already covered by routine nuclear safeguards procedures as carried out by IAEA and EURATOM. Current and future nuclear safeguards would be effective in finding anomalies for most scenarios of tritium production and would even suffice for tritium control in 20% of all paths. In other cases, additional measures (e.g., nondestructive analysis to identify lithium-6 or tritium accountancy) may be introduced which in turn may enhance the efficiency of nuclear safeguards.

Unreported breeding of tritium up to 100 g/y would be detectable by those nuclear safeguards activities which are already implemented by the IAEA in order to detect unreported breeding of plutonium from natural uranium. All neutron sources in which more than one significant quantity (8 kg) of plutonium can be produced during one year are under nuclear safeguards. Since tritium production is always in competition with plutonium production, all facilities and possible paths which can breed 110 g/y (i.e., the tritium equivalent of 8 kg plutonium per year) are already under nuclear IAEA safeguards.

In addition to evaluating routine safeguards, technologies which are used to verify the total uranium content and its enrichment, tritium control relies mainly on nondestructive detection of raw materials like lithium-6.

The insection of specially designed breeding targets into fresh fuel elements can be observed in fuel fabrication facilities (type 2). When this process is completed, the fuel elements are sealed. Further safeguards are carried out by item counting and seal inspection as well as by containment and surveillance measures which are already current practice.

Control procedures of **type II (verification of nonremoval)** to detect unreported removal of tritiated water are required at heavy water reactors (types 1b,d), spent fuel storages (type 3) and detritiation facilities (type 6). These can partly rely on IAEA safeguards required for heavy water which, according to their stated goal, would already detect the diversion of moderately tritiated heavy water containing more than 100 g of tritium within one year. The significant quantity of heavy water for nuclear safeguards is 20 t. Tritiated heavy water has a concentration of up to about 2 TBq/kg, i.e., 20 t contain some 110 g, the significant quantity of heavy water might have to be reduced with regard to tritium control. At present, safeguards on heavy water are implemented in Argentina only.

Table 3.5 Survey of acquisition paths, facilities concerned, and appropriate control activities for an agreement with $SQ/t_d > 1$ g/y.

	path [a]		max. div. rate [g/y] [b]	facility types	conditions on facilities	main control activities [c]
Li	1.	(a)	5,000	1e	shutdown	NTM
		(b)	1,300–2,000	1,2		IAEA, reactor power monitoring
	2.	(a)	2,200	1		IAEA
		(b)	1,000–5,000	1		IAEA
	3.	(a)	30	1a,c,2,3		IAEA, NDA-Li on fuel assemblies
		(b)	70–300	1a,c		IAEA, NDA-Li on control rods
		(c)	2–26	1a,c,2	PWR	IAEA
		(d)	30	1a,c	PWR	IAEA
		(e)	<30	1a,c	empty spaces outside core	IAEA
	4.		>1	1c,d	>200 kW_{th}	IAEA, NDA-Li on targets
	5.	(a–c)	100–500	1e	limited tritium breeding	^{6}Li acc. + C/S of targets, (Stern, 1988)
		(d)	<1	1c,d	fusion blanket research	– none –
	6.		10,000	1f	spallation neutron source	IAEA, NDA-Li on target
	7.	(a)	>1,000	1	seed-blanket reactor	future IAEA
		(b)	1,000–9,000	1	LMFBR (fast breeder)	Li in coolant
		(c)	>10,000	1f	proposed fusion reactor	future IAEA, tritium accountancy
		(d)	>10,000	1	fusion/fission hybrid	future IAEA, tritium accountancy
He	1.		≫1	1	gas cooled	^{3}He in gas
	2.		100	1	water cooled or moderated	^{3}He in water
	3.		10	1c,d	in case of RPE experiment	tritium accountancy
	4.		>1	1		IAEA
				1	if gas loop installed	^{3}He in loop
				1	if > 1g ^{3}He in loop	tritium accountancy
	5.		400	1		IAEA, NDA-^{3}He in target
	6.		≫1	1f	fusion neutron source	future IAEA, tritium accountancy
B	1.		0.03–0.6	1,2	excluding PWRs and FBRs	check boron control rods

Table 3.5 *(continued)*.

path [a]			max. div. rate [g/y] [b]	facility types	conditions on facilities	main control activities [c]
	2.		0.02–0.35	1	PWR	– none –
	3.		0.07	1	BWR	– none –
D_2O	1.	(a)	2,000	1b,d	>12 MW$_{th}$, if loss >150 kg	loss confirmation
		(b)	3	1b,d	>12 MW$_{th}$	monitor D_2O emission
		(c)	30	1b,d	>12 MW$_{th}$	IAEA (D_2O acc. and C/S)
	2.		$\ll$0.1	1	PWRs	– none –
	3.		>1	4,9	cap. > 240 t/y	C/S of aqueous HLW
	4.	(a)	3,000	6		H-3 concent., D_2O account.
		(b)	>1	6		tritium accountancy
fuel	1.		>1	4	cap. > 200 t/y	monitor tritium in off-gas
	2.		>1	4	cap. > 240 t/y	check tritium in aqueous HLW
	3.		>1	4	cap. > 170 t/y	C/S of organic waste
	4.		>1	4	cap. > 240 t/y	C/S of cladding waste
	5.		>1	4	cap. > 15 t/y	monitor tritium in aqueous emissions
	6.		>1	4,3	cap. > 12 t/y	IAEA
	7.		>1	4,3	cap. > 12 t/y	check fuel integrity
stor.	1.–6.		>1	7,8	Max(inventory,throughput) > 1 g	tritium accountancy
recov.	1.		>1	9		C/S of waste
	2.		>1	6,7,8	Max(inventory,throughput) > 100 g	monitor tritium in aqueous emissions
	3.		>1	6,7,8	Max(inventory,throughput) > 100 g	monitor tritium in off-gas
	4.		0.002	9	DD fusion experiments	– none –
	5.		>1	8		tritium accountancy
	6.		$\simeq$1	6,7,8	Max(inventory,throughput) > 0.001 g	tritium accountancy

[a] Numbering as in the diversion path analysis, Chapter 2.

[b] If dependent on reactor power, the figures are given per GW$_{th}$y. See Section 2.4.

[c] Abbreviations: NTM = national technical means; IAEA = nuclear safeguards as currently performed by IAEA and EURATOM; NDA-Li = nondestructive analysis to identify lithium-6; C/S = containment and surveillance.

Other control procedures of type II (verification of nonremoval) are required to detect unreported removal of spent fuel from reactors (types 1a–d) or storages (type 3). Current nuclear safeguards are designed to detect the diversion of fuel containing more than 8 kg of plutonium within 3 months. This corresponds roughly to 800 kg of spent fuel, which could contain some 60 to 180 mg of tritium depending on the fuel type and burnup. Thus, the illegal removal of spent fuel containing more than 0.2 g of tritium would be detected with existing control measures.

Further control procedures of type II (verification of nonremoval) have to be introduced at all tritium bulk-handling facilities (types 4 through 8) to verify nonremoval of tritium in various chemical forms. Tritium accountancy complemented by containment and surveillance are the appropriate control methods. Tritium accountancy is current practice at all tritium-handling facilities mainly for the purpose of radiation protection carried out by plant operators and verified by national authorities. Although there are specific technical problems because of tritium being a gas, it can be shown that an accountancy capability can be achieved which compares well with the capability of nuclear safeguards. The precedent for this type is due to an agreement between EURATOM and Canada.

Taking into account the existing tritium stocks and sources (see Table 2.5), the conclusion can be drawn that verification of nondiversion of tritium is feasible at reasonable costs. Not all facilities have to be inspected for tritium controls. Procedures to verify the nondiversion of tritium have to be introduced at a limited number of facilities (up to some 50, depending on the specific control instrument and membership) in which no nuclear materials except tritium are handled (see Table 3.4). There are fewer significant quantities of tritium to be controlled than there are significant quantities of plutonium under nuclear safeguards (see Table 3.1).

Other than the weak points of accountancy as identified and discussed in Section 3.4.5, there are no additional weak points for tritium control. The accountancy weakness can be managed by implementing additional control measures.

Physical protection of tritium cannot be guaranteed unless access to tritium is successfully prohibited, because it is easy to pass a tritium container through a check-out point without being detected. But any significant diversion of tritium can be detected within one year by tritium accountancy.

Consequently, from the technical perspective there are no fundamental problems regarding the introduction of tritium control procedures with state-of-the-art technology even if the significant quantity is very conservatively set at 1 gram.

3.9 Endnotes

1. For simplicity, safeguard systems applied to fissile materials, especially plutonium and highly enriched uranium, are in this study referred to as *nuclear safeguards*.

2. In fact, though intended to be an internal regime with conditionally open membership and striving for complete global participation, the nuclear nonproliferation regime has characteristics of an external regime, because potentially proliferating countries are nonmembers of the NPT or other instruments

of the control regime. Therefore, an additional goal of an international tritium control might be to discover whether de facto nuclear weapons states not members of these agreements receive enough tritium for them to move on to produce an arsenal of tritium-consuming nuclear weapons.

3. All terms used here are in accordance with the IAEA Safeguards Glossary (IAEA, 1987). Comprehensive descriptions of the theory and applications of effectiveness of control can be found in Avenhaus (1977, 1986).

4. Significant quantities for nuclear materials are: 8 kg for plutonium containing less than 80% ^{238}Pu, 8 kg for ^{233}U, 25 kg for uranium enriched to $\geq$20% of ^{235}U, 75 kg for uranium containing $<$20% ^{235}U, and 20 t for thorium (IAEA, 1987). The SQ for heavy water (D_2O) is 20 t, which is enough to achieve criticality in a small heavy water power reactor capable of producing one significant quantity of plutonium annually (Morsy, 1987).

5. The critical mass of a fissionable material that will just maintain a fission chain reaction under precisely specified conditions, such as the nature of the material and its purity, the nature and thickness of the tamper, the density after compression, and the physical shape.

6. See Colschen (1995). The differences in possible agreements depend on the goals they serve. For a list of possible goals see Section 1.5.

7. Quotation from Stern (1988).

8. Stern (1988) assumed 4 g of tritium per warhead. In this study 2 to 3 g are assumed; see Section 1.3.2.

9. In the case of minimum deterrence the excess number of weapons that would result in a counterforce capability is considered by Stern to be significant rather than an incremental increase of the stockpile of a few percent as in the case of large stockpiles. This figure will definitely provoke considerable debate as it deals with the number of warheads required for minimum deterrence.

10. See discussion of an integrated cutoff in Section 1.8.3.

11. According to the radiation protection ordnance of the Federal Republic of Germany (Strahlenschutzverordnung, §52 and App. IV), the derived dose limit for inhalation by an occupationally radiation-exposed person of category A implies that he or she is permitted to inhale no more than 2.2×10^8 Bq of tritium in the form of tritiated water during a three-month period. Assuming a sudden spill of 0.1 mg (360×10^8 Bq) of tritium at the beginning of an eight-hour working day, which will be homogeneously dispersed within a volume of 1000 m^3 (i.e., a concentration of 0.36×10^8 Bq/m^3), and furthermore assuming an inhalation rate of 0.85 m^3/h, the employee will, during this day, inhale a total of 2.5×10^8 Bq. This would exceed the derived dose limit. This rough estimate would have to be modified accordingly if air detritiation or ventilation systems were in operation.

    The air monitoring inside the Tritium Laboratory Karlsruhe will be designed to ensure that the daily allowable intake results in 3.6×10^6 Bq HTO, corresponding to 15 mSv per year under the assumption of 250 working days per year. The trigger level for alarm is set at 1×10^8 Bq/m^3 (Dilger, 1989).

12. The same is envisaged for two large European tritium-handling facilities Kraemer (1993) and Housiadas, Perujo and Vassallo (1993).

13. At present, amongst others, the following timeliness guidelines are used: 1 month for fresh fuel containing HEU, Pu, or MOX; 3 months for irradiated fuel containing HEU or Pu; and 12 months for fresh fuel containing natural uranium, LEU, or Th (IAEA, 1987). It is 12 months for D_2O (Baeckmann and Rosenthal, 1987).

14. In Stern (1988) a time of seven days is given for a certain diversion path. Stern calculates that a 2400 MW_{th} production reactor which was supposed to be kept on standby could produce—using its full capacity—an excess amount of 100 g (defined as one SQ) of tritium in approximately three days. Furthermore he assumes that roughly four more days are required to get the targets out of the reactor, extract the gas from the targets, purify and enrich the gas, and deploy the warheads which were completed by the supply of tritium gas. Hence, $t_c = 7$ d. It should be noted that it does not make much sense to unload the complete reactor core only a few days after start-up since each target would be burned by less than 1%. To be efficient the targets have to be kept in the reactor at least for some 200 days to achieve burnup in the order of 10%.

15. See remark above with explanation of $t_c = 7$ d and see Stern (1988).

16. Quotation from Stern (1988), p. 79.

17. See, for example, von Hippel *et al.* (1985), von Hippel and Levi (1986), Weinstock and Fainberg (1986), Thompson (1990), and Jasani and Stein (2002).

18. In 1969, the U.S. dropped the demand for inspection teams searching the country for clandestine facilities. Instead it announced that IAEA inspections at declared facilities would be considered adequate. The reason for this shift was that the development of surveillance satellites had made such progress that the U.S. was confident that they would be able to detect clandestine facilities by national technical means.

19. See, for example, Richter (1986), and von Hippel and Levi (1986).

20. The NCC is sometimes referred to as a uranium neutron collar. It is a transportable active-neutron interrogation system for LWR fuel assemblies which determines the fissile content. Fission is induced by subthreshold thermal and epithermal neutrons. The yield is a function of the fissile mass in the fuel assembly. ^{3}He-based detectors measure the relative influence of the induced fission neutrons from the fuel assembly.

21. A precedent for this is given in Argentina (Morsy, 1987).

22. Possible increases in inventory: measured input transfers, correction to receipts, nuclear production in a reactor, as appropriate.

23. Possible decreases in inventory: measured output transfers, measured or estimated accidental loss, radioactive decay, measured discard as waste, exemptions, as appropriate.

24. This term is used in IAEA safeguards systems accountancy. Other sources use the technical term "inventory difference" (ID), e.g., Ellefson and Gill (1986), Gill *et al.* (1986), and Wall and Cruz (1985).

25. Not to be confused with the expected accountancy capability E or the expected measurement accuracy for closing a material balance δ_E as defined below. E_1 covers the contribution to δ_E which is due to the measurement instruments and procedures. L and ΔH describe the contribution to δ_E which is due to the uncertain availability of tritium for measurements.

26. The threshold S expresses the accountancy verification goal and may be equal to the significant quantity (see Section 3.2.2).

27. The algorithm adopted by EURATOM is slightly different: The sealed inventory is excluded from determination of A; δ_E is replaced by δ, which is a weighted average uncertainty; and the numerical factor 3.29 is replaced by 1.65. The EURATOM goal is stricter than that of the IAEA.

28. See Morsy (1987). There are, however, no reference values which are considered achievable in practice for the measurement uncertainties expected for closing a material balance for heavy water.

29. The UWMAK reference fusion reactor design was developed by the University of Wisconsin Fusion Feasibility Study Group and corresponds to 5 GW thermal energy output.

30. General information on tritium measurement techniques can be found in National Council on Radiation Protection and Measurements (NCRP) (1976), Lindsay *et al.* (1987), Nickerson (1982, October), and Wood (1993).

31. The c in PVT/c stands for "concentration" measurement, which can be replaced by MS, GC, or BSD, whatever is appropriate.

32. The pressure necessary to tritiate uranium, for example, is higher than the decomposing pressure of uranium. Due to the lattice swelling that occurs during hydriding, the uranium breaks up into a fine powder. Thus, after a few cycles the initially solid getter bed is decomposed into a fine, porous powder which allows gas permeation. The consequences are an increased streaming resistance and the risk of uranium tritide transportation through the pipes.

33. See also subsection on discussions in the U.S. in Section 1.4.2.

34. The fraction of tritium that finds its way into the aqueous phase depends on the plant design. If the plant is working on the PUREX process and no tritium is extracted before the fuel is dissolved in acid, 80 to 85% of the tritium will be distributed in various aqueous waste streams. On the contrary, in the carbidic THOREX process for HTGRs, 90 to 95% of the tritium ends up in gaseous emissions (Schnez, 1975).

35. One example for this problem is the Tritium Aqueous Waste Recovery System (TAWRS) at Mound. Calculations have shown that with a constant tritium concentration in the feed stream, it could take 6 to 9 months before equilibrium concentration profiles are established again. Shutting down operations every 3 to 6 months would never allow the system to come to equilibrium for required operating conditions (Sienkiewicz, 1988).

36. The average percent difference was defined as $100 \times (P - A)/((P + A)/2)$, where P is the predicted inventory and A the actual one.

37. The following examples are given in Kapishev (1989): (1) Turbo molecular pumps will contain 4 to 60 TBq (0.01 to 0.17 g) each. Disposal is expected for a reference fusion reactor once a year. (2) Metals and alloys (stainless steel getter forming metals) contain 3.7 TBq/kg (0.01 g/kg). Disposal of 12 kg is expected several times during a 10-year period of a fusion reactor. (3) Zeolites from molecular sieves will contain 3.7PBq/kg (10 g/kg). Disposal of 20 kg is expected several times during a 10-year period of a fusion reactor.

38. This term is used in this monograph as it is defined in Wall and Cruz (1985).

39. A study carried out at Mound Laboratory revealed that total unmeasured transfers from sampling typically represent 50% of the total tritium gained in the Tritium Effluent Removal Systems (TERS), which capture all of the tritium from sampling. The remaining 50% is due to measured transfers and leakage from process to vacuum systems and glove boxes. The unmeasured quantity of gas removed by sampling was both calculated theoretically and determined experimentally. While the volume and number of samples can be known with fairly high accuracy, the actual flushing time and the flow rate are rather uncertain (Ellefson and Gill, 1986).

40. This volume is chosen so as to make a conservative assessment. The unmeasured transfer of gas per sampling of the process at Mound Laboratory has a sampling volume V_s of 10 to 50 cm^3 (STP) with a relative error varying from 20 to 70% ($2\,\sigma$) depending on the capillary used. This corresponds to 2.7 to 13.5 mg or some 1 to 5 TBq of T_2.

    In Sandia National Laboratories Livermore (SNLL), approved mass spectrometer sample containers may contain up to 1.8 TBq (5mg) of tritium and may be removed and transported for analyses without recording the transfer (Wall and Cruz, 1985).

    Tanase *et al.* took samples for gas chromatography with $V_s = (3.35 \pm 25)$ cm^3 (STP) (Tanase *et al.*, 1988).

    The method described by Ellefson earlier makes use of gas samples with $V_s = 2$ cm^3 (Ellefson, 1982).

41. Two examples are reported in Sauermann *et al.* (1979): A target with 3 TBq (0.008 g) lost 1.47 TBq within 5 months of operation; the next target with the same activity lost 1.58 TBq within 4 months.

42. From the volume rate and average tritium concentration the total stack emission of tritium can be determined. The volume rate is determined with a flow

meter. A by-path with about 1 m^3/h is used to condensate or freeze out the tritiated water. From time to time (e.g., every 3 months) a sample of the condensate is taken and analyzed for tritium in a scintillation detector. The lower detection limit is in the order of 0.01 Bq/m^3, which is about one order of magnitude lower than the normal tritium concentration in air (Gebauer, 1977). Liquid effluents can be monitored in a similar manner.

43. In the U.S., the most frequently used material for the containment of tritium is stainless steel SS316L (German norm 1.4404, similar to 1.4541 and 1.4301).

44. Genty *et al.* pointed out that the presence of grease, especially on ground sockets, causes a drastic decline in the tritium content in the long run (Genty *et al.*, 1973).

45. The reference tritium-handling system consists of a transfer line, storage tank, uranium getter pump with a design capacity of 1000 L(STP), molecular sieve, and a vacuum pump.

46. A method for measuring the quantity of hydrogen remaining on the uranium getter pump after inventory regeneration is presented in Ellefson and Gill (1986).

47. In the presence of oxygen contamination in a high-purity tritium gas, tritiated water vapor is readily formed and absorbed on the molecular sieves. Because of the affinity of zeolite for ammonia and the high exchange rate of hydrogen isotopes between ammonia and water, tritium arriving on zeolite as ammonia (NT_3) could well remain as water (HTO). Radiation-catalyzed direct exchange of tritium gas (T_2) with the water "pool" held in the molecular sieve is another possible mechanism for the incorporation of tritium into zeolite. This exchange is expected to be slow, but can be significant at elevated temperature (Carter *et al.*, 1964). Thus, the zeolite material holds accountable material.

48. See Gill, Ellefson, and Rutherford (1986). Both methods were applied with 10 to 20% agreement to five nonremovable zeolite traps in the Mound Laboratory. These traps had been subjected to tritiated water and ammonia vapors for approximately 4 years. Each trap contained 1.7 kg of zeolite (dry weight). "Normal" regeneration by evacuating the traps at $\simeq$250° C for 4 hours provided little recovery of tritium inventory. After heating the five traps for 16 hours at 400° C, the helium carrier gas removed a total of 258 g with a tritium mass gain of 12.5 g. It was estimated that after this (75±20) g of water and (4.8±1.5) g of tritium remained on the five molecular sieves. This was confirmed by the isotopic exchange method (Gill *et al.*, 1986).

49. See Kraemer *et al.* (1993) and Housiadas *et al.* (1993). The latter estimates 3.5 g for ETHEL, which includes 1.3 g annual accumulation of solid waste. This should in fact be accounted for as measured output. It can be measured with an error contributing to E_1 (see Equation 3.2).

50. See Section 2.6 for a survey of appropriate facilities and Section 3.7 for the inspection procedures.

51. See diversion path analyses, Section 2.5.

52. Examples of publicly known large accidental releases: Lawrence Livermore National Laboratory (LLNL), accidental release on August 6, 1970: 1.07×10^{16} Bq (29 g) (Myers *et al.*, 1973). Savannah River Plant, accidental release on May 2, 1974: 1.77×10^{16} Bq (48 g) (Quigg, 1984). NPD nuclear power plant at Rolphton, Canada, during the week of August 20, 1981: 130 TBq (360 mg) (Edwards, 1981).

53. Myers *et al.* reported results of measurements made after an accidental release of 1.07×10^{16} Bq through the stack of the Gaseous Chemistry Building of LLNL. All samples of water, milk, and urine contained normal background levels of tritium. Detectable levels of activity were found in some vegetation and atmospheric water vapor samples. The maximum values obtained from on-site samples were 440 Bq/m^3 for water vapor and 44 kBq/l for water in vegetation. Pre-accident levels were below the detection limit for water vapor ($<1\times10^{-3}$ Bq/l) and 400 Bq/l on average, 9 kBq/l at maximum, for vegetation (Myers *et al.*, 1973).

54. See above and Table 3.1.

55. A precedent for safeguarding heavy water is given in Argentina (Morsy *et al.*, 1987).

56. This diagram is taken from Kalinowski (1993). In this figure, a survey is given on the tritium inventories and predicted emission values as calculated by different studies on the safety of fusion energy (see x in Figure 3.2). For comparison and to broaden the view, empirical (∗) and calculated results (x) related to other tritium-handling facilities are included in the figure. The diagonal indicates the relative loss of 1 ppt per year (10^{-3}/a) with respect to the total inventory.

    This graphical representation should not imply a functional relationship between inventory and emissions. If a functional relationship were assumed, it would be more appropriate to use the maximum of inventory and throughput as the independent variable and take maintenance operations and equipment failures into consideration. It can be used to identify a qualitative pattern which shows different types of facilities lying in different regions of the inventory emission rate plane.

    All empirically based values (indicated by an ∗ in Figure 3.2) for relative emissions from facilities containing more than 10^4 TBq (28 g) lie above the 1 ppt diagonal — with one remarkable exception: Emissions from the Tritium Systems Test Assembly (TSTA) at Los Alamos National Laboratory achieved relative emissions as low as $1\text{-}3\times10^{-5}$ with fairly large inventories. Four points are given in Figure 3.2 for each historical phase during the stepwise increase of the inventory of TSTA (11, 30, 50, and 110 g).

57. In Millan and Chung (1977), the detection of a SO_2 plume 400 km from its source is described.

58. Currently, three facilities are subject to inspection due to the agreement between EURATOM and Canada: ETHEL, JET, and TLK (see Section 1.4).

59. For example, the French National Agency for Radioactive Waste Management (ANGRA) registers and labels each waste package individually. Through a bar code system with laser readers each package is monitored and its destination checked at every stage of handling. The quantities of each radionuclide are accounted by means of a central computerized system.

60. This kind of inspection will be started by EURATOM in the near future according to a bilateral agreement between EURATOM and Canada. The real inspection activities have not been published and may differ from those suggested here.

61. The larger of inventory and throughput applies.

62. This is already practiced, e.g., by ORNL. Within ten years they collected rejected light sources containing about 12% of the total amount of tritium put in such light sources by ORNL (Kobisk *et al.*, 1989). Some countries have already included provisions for the return or controlled discard of radioisotopes in licensed consumer products in their radiation protection legislation.

References

Arms Control Reporter (1989) Section 611.B.593. With reference to NYT 28.10.89.

Avenhaus, R. (1977) *Material Accountability — Theory, Verification, and Applications.* London.

Avenhaus, R. (1986) *Safeguard-Systems Analysis. With Applications to Nuclear Material Safeguards and Other Inspection Problems.* New York.

Avenhaus, R. and Spannagel, G. (1988) Analysis of Tritium Laboratory Accountancy Data. *Fusion Technology*, 14, 1102–1107.

von Baeckmann, A. and Rosenthal, M.D. (1987) Selection of a Safeguards Approach for the Arroyito Heavy Water Production Plant. IAEA-SM-293/140. In IAEA (1987). *Nuclear Safekaliards Technology 1986*, Proc. Intern. Symp. on Nuclear Materials Safeguards, November 10–14, 1986, Vienna.

Banner, A.V., Young, A.J. and Hall, K.W. (1990) Aerial Reconnaissance for Verification of Arms Limitation Agreements. An Introduction. UNIDIR/90/83, New York.

Barnaby, F. (ed) (1990) *A Handbook of Verification Procedures.* New York.

Broad, W.J. (1989) U.S. halts sale of tritium after loss of enough to make a nuclear bomb. *The New York Times*, July 26.

Buchberger, T., Rauch, H. and Seidl, E. (1989) Tritium and helium-3 in metals investigated by neutron radiography. *Kerntechnik*, 53, 215.

Canadian Fusion Fuels Technology Project (CFFTP) (1988) FLOSHEET — A Process Simulation Program for Modelling Fusion Fuel and Hydrogen Isotope Processing Systems, *Product Bulletin*, 4/88.

Carter, J.L., Lucchesi, P.J. and Yates, D.J.C. (1964) *J. Phys. Chem.*, 68, 1385.

Canty, M.J. (1985) Comments on Inspection Goal Criteria for Material Accountancy. JOPAG 09.85-PRG-119, October, Jülich.

Cheek, C.H. and Linnenbom, V.J. (1958) The Radiation-induced Formation of Ammonia, *J. Phys. Chem.*, 62, 1475.

Cochran, T.B., *et al.* (1987) U.S. Nuclear Warhead Production. In *Nuclear Weapons Databook*, Vol. 2. Cambridge.

Colschen, L.C. (1995) Die Internationalisierung der Tritiumkontrolle als Baustein des Nichtweiterverbreitungsregimes für Kernwaffen. Bedingungen, Möglichkeiten und Chancen. Ph.D. thesis to be submitted to Free University Berlin.

Colschen, L.C., Kalinowski, M.B. (1994) Can International Safeguards be Expanded to Cover Tritium? Paper IAEA-SM-333/27, In *Proc. IAEA Symposium on "International Nuclear Safeguards 1994: Vision for the Future,"* March 14–18, Vienna. Proceedings Series No. 945, Vol. 1, 493–503.

Combs, F. and Doda, R.J. (1979) Large-Scale Distribution of Tritium in a Commercial Product. IAEA-SM-232/57. In *Behaviour of Tritium in the Environment. Proc. Symp.*, San Francisco, October 16–20, 1978, International Atomic Energy Agency, IAEA-SM-232/49, Vienna.

Dilger, H. (1989) Strahlenschutzaspekte beim Tritiumlabor Karlsruhe (TLK). In *Jahresbericht 1988 der Hauptabteilung Sicherheit, Kernforschungszentrum Karlsruhe*, KfK-4530, Karlsruhe.

Edwards, G. (1981) Background Information on the spill of 3500 Ci of tritium into the Ottawa River from NPD nuclear power plant at Rolphton during the week of August 20.

Ellefson, R.E. (1982) Process Monitoring of Tritium Concentration. Monsanto Research Corporation — Mound, MLM-2995-OP, Miamisburg.

Ellefson, R.E. and Gill, J.T. (1986) Tritium inventory differences. I. Sampling and U-getter pump holdup, Monsanto Research Corporation — Mound, MLM-3369-OP.

Gabowitsch, E. and Spannagel, G. (1989) Modeling and simulation of tritium handling system, *Kerntechnik*, 53, 202.

Gebauer, H. (1977) Meßtechnische Erfassung radioaktiver Ableitungen aus Kernkraftwerken, *Atomwirtschaft*, 22, 149–153.

Genty, C., *et al.* (1973) Determination of tritium in an analytical chemistry laboratory. In *Tritium Conf. Proc.*, A.A. Moghissi, M.W. Carter (eds), August 30–September 2, 1971, Las Vegas, Nevada.

Gill, J.T., Ellefson, R.E. and Rutherford, W.M. (1986) Tritium inventory differences. II. Molecular sieve holdup. Monsanto Research Corporation — Mound. MLM-3370-OP.

Gupta, V. (1992) Algeria's Nuclear Ambitions. *Jane's International Defence Review*, No. 4, pp. 329–330.

Hessisches Ministerium für Umwelt (HMU) (1983) Verfahren zur Bestimmung von Tritium in Oberflächenwasser. B-H-3-OWASS-01-01, Wiesbaden.

von Hippel, F., Albright, D. and Levi, B. (1985) Stopping the Production of Fissile Materials for Weapons, *Scientific American*, 253, No. 3, 26.

von Hippel, F. and Levi, B.G. (1986) Controlling Nuclear Weapons at the Source: Verification of a Cutoff in the Production of Plutonium and Highly Enriched

Uranium for Nuclear Weapons. In *Arms Control Verification: The Technologies That Make it Possible*, K. Tsipis, D.W. Hafemeister, P. Janeway (eds). Pergamon-Brassey, Elmsford, NY.

Housiadas, C., Perujo, A. and Vassallo, G. (1994) The Control of Tritium in ETHEL, *J. Fusion Energy*, 13, 455–460.

Hugony, P., Sauvage, H. and Roth, E. (1973) La production de tritium en France, *Bulletin d'Information Scientifique et Technique*, No. 178, February, 3–17.

IAEA (1987a) IAEA Safeguards Glossary. Vienna.

IAEA (1987b) Nuclear Safeguards Technology 1986. In *Proc. intern. symp. on nuclear materials safeguards*, November 10–14, 1986, Vienna.

IAEA (1993) The Annual Report for 1992. Vienna.

Jasani, B. and Stein, G. (2002) (eds) *Commercial Satellite Imagery. A Tactic in Nuclear Weapon Deterrence*. Springer-Verlag, Berlin etc.

Kalinowski, M.B. (1993) Uncertainty and Range of Alternative in Estimating Tritium Emissions from Proposed Fusion Power Reactors and their Radiological Impact, *Journal of Fusion Energy*, 12, 157–161.

Kalinowski, M.B. (1997) *Monte Carlo Simulationen und Experimente zum zerstörungsfreien Nachweis von Lithium-6. Physikalische Fragen zur Tritiumkontrolle*. Shaker Verlag, Aachen.

Kapishev, V. (1989) Tritium Radiation Safety of Fusion Installations and Reactors. In *Proc. Course and Workshop on Tritium and Advanced Fuels in Fusion Reactors*, International School of Plasma Physics 'Piero Caldirola,' Varenna, September 6–15.

Kobisk, E.H., *et al.* (1989) Tritium-Processing Operations at the Oak Ridge National Laboratory with Emphasis on Safe-Handling Practises, *Nucl. Instr. Methods*, A282, 329–340.

Kraemer, R., *et al.* (1993) Common Tritium Control Methodology Proposed for Two Civil Tritium Facilities ETHEL and TLK. In *15th Annual Meeting of ESARDA*, May 11–13, 1993, Rome.

Kurz, U. (1984) Der U/UT_3-Lagerbehälter im Tritium-Lager. Jül-Spez-258, Jülich.

Lindsay, C.N., Sprague, R.E. and Brandenburg, J.A. (1987) A Measurement Control Study for Tritium Gas. MLM-3441, Mound Laboratory.

Lu, M.-S., Zhu, R.-B. and Todosow, M. (1988) Unreported Plutonium Production in Light Water Reactors. ISPO-282, TSO-88-1, February, Brookhaven National Laboratory.

Luthardt, G. and Huschka, H. (1983) Erarbeitung von sicherheitsrelevanten Daten zur industriellen Entsorgung tritiumhaltiger Abwässer auf der Basis großtechnischer Untersuchungen. Bericht BMI-1983-006.

Millan, M.M. and Chung, Y.S. (1977) Detection of a Plume 400 km from the Source, *Atmospheric Environment*, 11, 939–944.

Miller, J. (1993) Tritium Accounting by Calorimetry. *Fusion Canada*, Issue 22, October.

Moghissi, A.A. and Carter, M.W. (eds) (1973) *Tritium*. Phoenix and Las Vegas.

Morsy, S., *et al.* (1987) Evaluation of Heavy Water (D_2O) Material Balance in Power Reactors. IAEA-SM-293/15. In *IAEA. Nuclear Safeguards Technology 1986*, Proc. intern. symp. on nuclear materials safeguards, November 10–14, 1986, Vienna.

Myers, D.S., Tinney, J.F. and Gudiksen, P.H. (1973) Health Physics Aspects of a Large Accidental Tritium Release. In *Tritium Conf. Proc.*, A.A. Moghissi, M.W. Carter (eds), August 30–September 2, 1971, Las Vegas, Nevada.

National Council on Radiation Protection and Measurements (NCRP) (1976) Tritium Measurement Techniques. NCRP Report No. 47, Washington.

Nickerson, S.B. (1982) The Tritium Monitoring Requirements of Fusion and the Status of Research. CFFTP-G-F82003, October.

Quigg, C.T. (1984) Tritium warning. *The Bulletin of Atomic Scientists*, 56, March.

Richter, R., *et al.* (1986) Analysis of LANDSAT TM images of Chernobyl. *Int. J. Remote Sensing*, 7, 1859–67.

Sauermann, P.F., *et al.* (1979) Fifteen Years of Experience in Handling Tritium Problems in Connection with Low-Energy Particle Accelerators. In *Behaviour of Tritium in the Environment. Proc. Symp.*, San Francisco, October 16–20, 1978, International Atomic Energy Agency, IAEA-SM-232/49, Vienna.

Scheffran, J. (1989) Verification and Stability. The Strategic Impact of Uncertainty and Perceptions. In *6th International AFES-RESS Conference*, December 8–10, 1989, Mosbach-Neckarelz.

Schnez, H. (1975) Behandlung und Abtrennung der radioaktiven Spaltprodukte Tritium, Edelgase und Jod in Kernbrennstoffwiederaufarbeitungsanlagen. Jül-1223, Juli, Jülich.

Schröder, K.J.P. and Roether W. (1979) The Releases of Kr-85 and Tritium to the Environment and Tritium to Kr-85 Ratios as Source Indicators. In *Isotope Ratios as Pollutant Source and Behaviour Indicators*, IAEA, Vienna.

Sienkiewicz, C.J., *et al.* (1988) Physical Inventory by Use of Modelling for the Tritium Aqueous Waste Recovery System, *Fusion Technology*, 14, 1096.

Stern, W.M. (1988) Nuclear Weapons Material Control: Verification of Tritium Production Limitations. Master thesis, MIT, Cambridge.

Tanase, M., *et al.* (1988) Production of 40 TBq Tritium Using Neutron-Irradiated ^{6}Li-Al Alloy, *Journal of Nuclear Science and Technology*, 25, 198.

Thompson, G. (1990) Verification of a Cut-Off in the Production of Fissile Material. In *A Handbook of Verification Procedures*, F. Barnaby (ed.), New York.

Tritium Test Facility (1982) *Physics Today*, November, 18.

Tsipis, K., Hafemeister, D.W. and Janeway, P. (1986) *Arms Control Verification. The Technologies That Make it Possible.* Pergamon-Brassey, Elmsford, NY.

Vance, D.E., Smith M.E. and Waterbury, G.R. (1979) Quantitative Determination of Tritium in Metals and Oxides. LA-7716.

Varley, J., Dingle, A. and Gee, S.C. (1993) *World Nuclear Industry Handbook 1993.* Sutton.

Wall, W.R. and Cruz, S.L. (1985) Tritium Control and Accountability Instructions. SANDIA Report SAND 85-8227.

Weinstock, E. and Fainberg, A. (1986) Verifying a Fissile Material Production Freeze in Declared Facilities with Special Emphasis on Remote Monitoring. In *Arms Control Verification. The Technologies That Make it Possible*, K. Tsipis, D.W. Hafemeister, P. Janeway (eds). Pergamon-Brassey, Elmsford, NY.

Weise, L. (1986) Bestimmung von effektiven Wirkungsquerschnitten und Erzeugungsraten für Tritium im Bestrahlungsexperiment TRIDEX. Jül-2058, April, Jülich.

Weiss, W., *et al.* (1986) Mesoscale Transport of Kr-85 Originating from European Source. *Nucl. Inst. Meth.*, B17, 571–574.

Wood, M.J., *et al.* (1993) Tritium Sampling and Measurement. *Health Physics*, 65, 610–27.

Chapter 4

Technical assessment of an international tritium control agreement

4.1 Adequacy and appropriateness

In contrast to fissile materials, tritium is not a necessary component for nuclear weapons, it plays a key role when advancing from simple fission devices to fusion boosted or thermonuclear weapons (see Section 1.3.2). Most nuclear weapons of all recognized nuclear weapon states are believed to contain tritium. There is evidence that most de facto nuclear weapon states and countries which are suspected of carrying out clandestine nuclear weapons programs seek to use tritium in their nuclear weapons (see Section 1.7). Tritium has strategic significance because warheads can be built to be smaller and lighter while retaining the same yield. There are strong arguments calling for an international tritium control (see Section 1.5).

Tritium control which is limited to the national level was found to be inadequate. In the past few years tritium control has been hesitantly expanded to the international level, but it still follows an incoherent and insufficient approach and most proliferation paths remain open (see Section 1.4).

International tritium control should be introduced in a way which is *adequate* with respect to its military significance. All verification procedures should be *appropriate* with respect to the detection goals.

In the case of tritium control, it is adequate to define the required verification goals less stringently than in nuclear safeguards because tritium is not necessary for the production of first-generation nuclear weapons.

In particular, the criterion timeliness (see Section 3.2.3) is much less significant as compared to nuclear safeguards. Since diverted tritium can be prepared for use in warheads in just a few days, the inspection frequency would need to exceed 52 times per year if the timeliness goal were defined as rigorously as in nuclear safeguards. This inspection frequency would be out of proportion because of the limited significance of tritium for the horizontal nonproliferation of nuclear weapons as compared to fissile materials. Moreover, it would not be practical in terms of required manpower due to the limitations set by the IAEA verification budget. Therefore, this study assumes that a desired *detection time* t_d, i.e., the maximum time span between diversion and its detection, of one year is appropriate and adequate.

One year appears *appropriate* because it is a natural time frame frequently used as the reference time in nuclear safeguards. The inspections would still fulfill a veritable deterrence function against diversion by bearing the risk of early detection.

At this point it becomes obvious that the *adequacy of verification* is very much a matter of political assessment and cannot only be measured in technical terms. If the verification goals are politically satisfying and acceptable, there will be no need to stretch technical verification procedures beyond this required level.

The *significant quantity (SQ)* (see Section 3.2.2) may also be defined less strictly than in nuclear safeguards. This definition is guided by the quantity required to build *one* crude nuclear weapon. The average amount of tritium used for one nuclear warhead is around 2 to 3 g. Therefore, a significant quantity of 10 g might be considered adequate because it is not plausible that only a single warhead is in place or supplied with tritium at the moment of introducing tritium into the nuclear arsenal. Nevertheless, the SQ is set to 1 g in this study as a rather conservative assumption, because it was possible to demonstrate the feasibility of verification even for such an extremely ambitious goal.

Unacceptable side effects of an international tritium control can be avoided. The tritium controls outlined in this study (see Chapter 3) can be performed without undue demands or unacceptable health risks to inspectors and operators (e.g., portable instruments, minimized radiation dose). They are therefore *appropriate* with respect to the goals.

4.2 Nondiscrimination

International tritium control should be implemented in a way that does not discriminate against any country or group of countries. There are no technical problems that would block or hamper the implementation of a nondiscriminatory tritium control in all countries, regardless of their nuclear weapons status. For political reasons, separate agreements to prevent the horizontal (see Section 1.7) and vertical proliferation (see Section 1.6) of tritium are likely, but they can be linked.

4.3 Feasibility and completeness

In order to find out what kind of verification activities would be needed, a comprehensive *diversion path analysis* was performed (see Chapter 2).

Diversion means the clandestine production of tritium or its illegal removal from existing tritium stocks for unknown and possibly nuclear weapons purposes. In nature, tritium occurs at concentrations which are far too low to make its extraction practically achievable. One reason for this is its comparatively rapid radioactive decay with a half-life of 12.3 years, i.e., a given quantity of tritium decreases at a rate of about 5.5% annually.

Since there are no exploitable natural sources, tritium has to be produced artificially by a nuclear reaction. Therefore, a diversion path analysis has to take into account possibilities of artificial tritium production as well as opportunities to remove tritium from existing stockpiles. Apart from actually performed production (deliberately or inadvertently), the potential production capacity is of interest for a diversion path analysis. According to the verification goals as defined above, all diversion paths by which more than 1 gram per year can be acquired are considered in this analysis. Altogether, 55 different paths are described, five of which do not exceed the critical production rate. A number of others rely on technologies such as fusion reactors, which do not yet exist but may be realized in the future.

Significant quantities can be achieved only by neutron capture and with a high neutron flux, as can be found in nuclear reactors. The *most relevant paths* to illegally produce pure tritium in significant quantities are those which are based on a nuclear reaction with a high cross-section and the possibility of inserting large quantities of the raw material into the flux of a nuclear reactor. These paths are *breeding of tritium from lithium-6* (see Section 2.4.1) *or helium-3* (see Section 2.4.2), *extraction of inadvertently produced tritium from heavy water* (see Section 2.4.4) *or ternary fission* (see Section 2.4.5), *and removal of stored pure or recoverable tritium* (see Section 2.5). The most efficient and economic way to produce tritium is the lithium path. This is pursued in all the recognized nuclear weapons states. Helium-3 targets are considered for future accelerator-based breeding systems. But this option, although technically feasible, is not currently being followed on a significant scale, because it is difficult to introduce large quantities of gaseous helium-3 into the reactor core.

Canada is the largest producer of tritium for civilian purposes. It extracts up to six times as much tritium from the heavy water moderator and coolant of CANDU power reactors annually as demanded worldwide for civilian purposes.

As a result of the comprehensive diversion path analysis it becomes clear that a nuclear reactor can be regarded as the bottleneck for today's tritium production. Accordingly, control measures would likewise be confined to nuclear reactors and to some additional related facility types such as fuel fabrication plants as well as any facilities containing significant quantities or throughputs of tritium. This indicates that tritium control is *in principle feasible.*

In order to prove that tritium control is *actually feasible,* appropriate verification procedures are listed (see Section 3). These can be classified in two categories: *Type I* is required to verify nonproduction of tritium (see Section 3.3) and covers deliberate breeding from lithium-6 or helium-3 as well as inadvertent production in heavy water and due to ternary fission. *Type II* is required to verify nonremoval (see Section 3.4) from existing stocks such as tritiated heavy water, spent fuel, or pure tritium gas.

As a result of the verification study, all diversion paths which could yield more than one significant quantity (1 g) of tritium within the desired detection time (1 year) could be covered by control measures (see Section 3.8). Any relevant diversion is detectable. Cheating or circumventing of these control measures is not easily possible and at least too expensive or risky to be of practical interest to the potential diverter. The applied measurement techniques and control procedures are *reliable and without loopholes.*

There are only a few *technical problems*, most of which seem to be solvable. The identified weak points of a tritium control that covers all 50 relevant diversion paths can be solved by adding the following control measures:

- In case of an anomaly in the results of routine nuclear safeguards applied to fresh nuclear fuel (see Chapter 3.3.3), technical means are required which allow the nondestructive identification of lithium-6 with a sufficiently high detection probability. It is has been shown that appropriate means can be developed. At least one of the 12 measurement principles reviewed in this study appears feasible: nuclear resonance absorption. In any case, destructive methods are readily available.

- The mathematical determination of the quantity of freshly produced tritium bears considerable uncertainties (see Section 3.4.2). Therefore, it has to be ensured by adequate containment and surveillance (see Section 3.5) that tritium is not removed illegally immediately after production and before the initial physical inventory is taken, which is the only reliable method to determine the baseline for accountancy.

- Accidental unrecorded emissions can be verified only qualitatively (see Section 3.4.4). Therefore, it has to be ensured that tritium is safely contained, significant quantities are distributed among multiple storages, emissions are monitored adequately, and the integrity of monitors is verified. These measures contribute to radiation protection as well. If need be, the concentration of tritium in the environment can serve as a rough check of tritium spills.

- The measurement of large amounts of tritium in waste (>5 g/y) might pose a problem for accountancy (see Section 3.4.4). Therefore, the amount of tritium in waste has to be minimized. This is welcomed for economic reasons as well.

- If the use of fusion energy was realized, which burns and produces huge amounts of tritium and releases large quantities of tritiated waste, it would be highly unlikely that the material unaccounted for (MUF) (see Section 3.4.4) can be kept below the significant quantity (1 g). In that case, containment and surveillance (see Section 3.5) of tritium inventories would play a major role.

Physical protection of tritium cannot be guaranteed unless access to tritium is successfully prohibited, because it is easy to pass a tritium container through a checkpoint without being detected. However, it should be noted that in the case of fissile materials, physical protection is not a goal of nuclear safeguards and any significant diversion of tritium can be detected within one inspection period by tritium accountancy.

Therefore, the *conclusion can be drawn that a complete coverage of all significant diversion paths can be achieved by tritium control.*

4.4 Control effectiveness

As described so far, comprehensive tritium control is in principle and actually feasible. It remains to be shown that it can be implemented effectively. The same standards as for nuclear safeguards are applied to tritium control in this study. However, the politically required verification effectiveness may be lower than in nuclear safeguards because tritium is not necessary for the production of first-generation nuclear weapons.

In order to assess the effectiveness, the inspection goal is described in quantitative terms. The objective of tritium control is timely detection (detection time t_d) of illegal removal of significant quantities (SQ) of tritium from peaceful activities to manufacture nuclear weapons or for purposes unknown with a certain detection probability $(1 - \beta)$ and the localization of tritium and satisfactory clarification of anomalies with a low risk of false alarm probability α (see Section 3.2).

The lower detection limit for raw materials such as lithium-6 should be less than the quantity required to produce one significant quantity (SQ) of tritium within the desired detection time t_d. In this study, SQ is defined conservatively to be one gram and t_d has been set to one year (see Section 3.2).

Although the required time to divert tritium and to use it in nuclear weapons may be much shorter than one year, it can be assumed that an inspection period of one year is adequate and appropriate. In addition, it can be asserted that this period of time is a good choice regarding effectiveness. Tritium control would be most effective if it were carried out simultaneously with nuclear safeguards. The latter are carried out with an inspection period of about one year or more frequently.

The *effectiveness of verification of nonproduction* (type I) is demonstrated in this study by using the example of the lithium path, which is the most important diversion path (see Section 2.8).

To verify the nonproduction of tritium, the technical means to produce one significant quantity of tritium have to be detected. Around 10 g of lithium-6 are required if 1 g tritium were to be produced within one year in a nuclear reactor (see Section 2.4.1). Monte Carlo simulations of the Neutron Coincidence Collar, which is routinely used for nuclear safeguards on fresh fuel assemblies, indicated that such amounts of lithium-6 added clandestinely to a fresh fuel assembly would cause a clearly detectable drop in the coincidence count rate and constitutes an anomaly to the measurements which would be detected by the inspectors (see Section 3.3.3). Therefore, a very high detection probability can be achieved. It has further been demonstrated that gadolinium, a possible cover-up material, causes a signal which is different from lithium-6, thus ensuring a low false alarm probability.

If an anomaly cannot be resolved, further, preferably nondestructive analysis is required. A review of 12 measurement principles based on active interrogation with neutrons or gamma rays indicated that a method is available which has a good selectivity (low false alarm probability). Nuclear resonance fluorescence (NRF) with gamma ray Bremsstrahlung spectrum at energies up to 5 MeV applied to a target containing lithium-6 would give a characteristic fluorescence gamma peak at 3.56 MeV, which is unique for this isotope. For a realistic scenario, a measurement time in the order of 10 minutes would suffice to detect 10 g of lithium-6 in one target rod. The required instrumentation is portable and suitable for routine in-field inspection (see Section 3.3.3). The interrogating gamma beam does not pose an unacceptable radiation hazard to personnel and does not cause significant activation or collateral damage to the inspected specimen.

In this study, the *effectiveness of verification of nonremoval* (type II) is evaluated quantitatively with respect to stocks of separated tritium and tritium in tritiated water. Verification of the nonremoval of tritium contained in spent fuel is as efficient as the nuclear safeguards on which it relies.

The accuracy of tritium inventory measurements in different modes (gaseous and aqueous, storages and operating systems) was extensively analyzed to estimate typical values of *material unaccounted for (MUF)* when closing a material balance (see Section 3.4).

Basically it turns out that tritium safeguards can be achieved with state-of-the-art technology. The expected detection capability in an imaginary closing of the material balance for the whole world inventory of all separated, civilian tritium stocks would amount to some 66 to 1300 SQ of tritium. For plutonium the respective number is 370 to 740 SQ_{Pu} of plutonium (1 SQ_{Pu} = 8 kg). Though this is criticized as being a poor performance, it proves that under normal operational conditions tritium accountancy is possible on a routine basis with an accountancy capability which compares well with the capability required for nuclear safeguards (see Section 3.4.5).

4.5 Minimum interferences with facility operation

Any disturbance of normal operation and any economic burden to the facility would be negligible because tritium control can in general either be integrated in already existing routine nuclear safeguards or in tritium accountancy for radiation protection purposes.

4.6 Minimum intrusiveness

Only information which is indispensable for control purposes needs to be obtained by inspectors and the controlling authorities. Data and qualitative information that is protected by patent or claimed as trade secrets are not required for control purposes. Information on tritium production considered sensitive to national security need not be revealed, since the basic knowledge on tritium production as published in unclassified literature is sufficient to design and apply an international tritium control and has been proven comprehensive enough to conclude this technical assessment of tritium control.

4.7 Synergies with other control procedures

It is easy to design control procedures for tritium in a way that makes optimum use of nuclear safeguards which are already in place, because most tritium control goals to verify nondiversion are already covered. IAEA-type safeguards would find anomalies for most scenarios of tritium production, and only a limited number of additional measures are necessary; nuclear safeguards can profit from some of these additional measures.

For example, the following diversion activities should be detectable by current nuclear safeguards with only minor supplementary measures, simply by additionally evaluating data and reports for tritium safeguards:

- Unreported breeding of tritium up to 100 g/y would be detected by those nuclear safeguards activities which are already implemented by the IAEA in order to detect unreported breeding of plutonium from natural uranium. All neutron sources in which more than one significant quantity (8 kg) of plutonium can be produced in one year are under nuclear safeguards. Since tritium production is always in competition with plutonium production, all facilities and possible paths which can breed 110 g/y (i.e., the tritium equivalent of 8 kg of plutonium per year) are already under nuclear IAEA safeguards, though the Agency is not allowed to examine tritium-related activities.

- The introduction of specially designed breeding targets into fresh fuel elements can be observed at fuel fabrication facilities. When this is accomplished, the fuel elements are sealed. Further routine safeguards are carried out by item counting and seal inspection as well as by containment and surveillance measures.

  The detection of tritium breeding targets in fuel elements can probably be achieved with routine measurement of the uranium enrichment and total fissile material content of fresh fuel assemblies after their final assembly. This is done with the Neutron Coincidence Collar (NCC), which would show anomalies on the insertion of undeclared target rods. For example, lithium-6 targets in fresh fuel elements would cause a drop in the count rate. The production scheme most difficult to detect would be the covering up of lithium in the fuel by declaring it to be gadolinium. Monte Carlo calculations indicate that even under unfavorable circumstances routine measurements with the Neutron Coincidence Collar would result in anomalies (see Section 3.3.3).

- Any targets inserted into the power reactor core outside fuel assemblies can be detected by visible inspection, video cameras, and other nondestructive safeguards applied by the IAEA at the reactor core. Therefore, this production path can be detected even if only a single small target with lithium-6 or helium-3 for the breeding of less than 1 g of tritium is placed in the reactor core.

- The diversion of safeguarded spent nuclear fuel containing more than 8 kg of plutonium should be detected within 3 months. This corresponds roughly to 800 kg of spent fuel, which contain some 60 to 180 mg of tritium depending on fuel type and burnup. Thus, the diversion of more than 0.2 g of tritium confined in spent fuel will be detected by nuclear safeguards within a quarter of a year.

- At present, safeguards for heavy water are implemented in Argentina only. The significant quantity of heavy water with respect to nuclear safeguards is 20 t. At a tritium concentration of about 0.02 TBq/kg, this quantity would contain 1.1 g. Therefore, the diversion of about 1 g of tritium contained in slightly tritiated heavy water (0.02 TBq/kg) could be detected by safeguards on heavy water. However, the tritium concentration in heavy water may be higher by more than two orders of magnitude.

The control tasks to verify nonremoval from existing stocks would have to be introduced at civilian tritium handling and production facilities with inventories or an annual throughput greater than 1 g. These tritium safeguards coincide with routine measures, which are taken for radiation protection purposes. At tritium handling facilities, accounting procedures established for radiation protection are far more precise than required for safeguards purposes. The data obtained from tritium accountancy would have to be made available to the IAEA. The agency could then evaluate and verify them with on site inspections.

The U.S. Department of Energy (DOE) has a system of accounting and control (described in Section 1.4). All tritium quantities equal to or greater than 0.0005 g (0.18 TBq) are considered accountable within one material balance area (MBA). For DOE reporting purposes, the accountable quantity of tritium for each facility is one order of magnitude larger but still much lower than required for effective verification of nonproliferation.

Other control procedures for nonproliferation and radiation protection are not in any way hampered or compromised by tritium control. On the contrary, nuclear safeguards can be improved by being supplemented with tritium control procedures. In any case, the control of tritium is of relevance to support horizontal nonproliferation and therefore would strengthen this regime.

4.8 Costs

The limitations set by the IAEA verification budget have to be taken into account. As far as reactors, fuel fabrication, and fuel and waste storage facilities are concerned, the appropriate solution is to tie tritium inspections to the routine IAEA inspection procedures for fissile materials. The number of inspection measures at these facilities in addition to nuclear safeguards can be kept small. However, all facilities with an inventory or production capacity of 1 gram or more per year should be inspected at least once per year. Thus, some tritium handling facilities in industry and research have to be inspected solely for tritium, but only a small number of facilities worldwide are affected. The exact number depends on the scope and membership of an international tritium control agreement. A survey of all relevant facilities worldwide revealed that a maximum of 50 facilities would have to be inspected in addition to the roughly 1000 facilities currently under nuclear safeguards (see Section 3.7).

The costs of a tritium control system can be minimized. No large investments for extra instruments are required for routine inspections. Instruments and equipment for efficient measurement and data handling are available at reasonable costs. Inspectors would need additional training and instruments which could be provided by national support programs.

Even if tritium control was to be established on its own and could not draw on nuclear safeguards, the required resources would be less than for nuclear safeguards because the worldwide amount of relevant material is smaller. There are less significant quantities of tritium to be controlled than there are significant quantities of plutonium under nuclear safeguards.

4.9 Effects on civilian tritium uses

Control procedures apply only to a very limited number of civilian tritium uses because quantities used in industry and research for civilian purposes are generally much smaller than the amount of tritium relevant for nuclear weapons. This will be true for approximately the next 50 years. After that time, fusion energy may be technically available for power supply.

The extra burden imposed by tritium control is not excessive, but still severe enough to encourage the conversion of applications to other isotopes, if possible.

4.10 Acceptability

Facility operators and countries would not find themselves at any substantial disadvantage by accepting international tritium controls. On the contrary, they could demonstrate their commitment to the goal of nonproliferation and avoid being suspected of keeping military options in mind.

4.11 Conclusions

A comprehensive and coherent international control system for tritium does not yet exist, but it is desirable to develop appropriate control instruments for reversing the vertical proliferation in recognized nuclear weapon states as well as to prevent the horizontal proliferation of advanced weapon designs into states with secret nuclear weapon programs. The nuclear nonproliferation regime would be strengthened by expanding its scope to tritium. An important precedent is given by the agreement between EURATOM and Canada to place tritium supplies from Canada under control.

For this purpose an *International Tritium Control System (ITCS)* as well as a verified cutoff for the production of fissile materials and tritium for military purposes, the *Integrated Cutoff (ICO)*, has been outlined. These two agreements would be complementary to each other. The goals are to avoid any undeclared diversion of tritium for military purposes, i.e., to block horizontal proliferation as well as to reduce — preferably to zero — the use of tritium for nuclear weapons purposes in all states possessing nuclear weapons so as to reverse vertical proliferation as a step towards complete nuclear disarmament.

This study raises the question of whether the necessary technical prerequisites for international tritium control are fulfilled. In particular, the technical feasibility of verification is assessed and compared to the standards agreed upon for nuclear safeguards. For the assessment, the significant quantity of tritium is assumed to be one gram,[1] and a detection time of one year should suffice.

The required verification measures as obtained by a comprehensive diversion path analysis entail two different control tasks. Both the *nonproduction* of fresh supplies and the *nonremoval* from existing and declared inventories of tritium have to be verified.

As a result of this study, verification appears to be technically feasible with state-of-the-art technology.

Verification of nonproduction can rely basically on current IAEA safeguards. It uses mainly detection and nondestructive analysis of possible irradiation targets at fuel fabrication plants and nuclear reactors to ensure the absence of lithium-6, the main raw material for tritium breeding. It was demonstrated with *Monte Carlo simulations* of the Neutron Coincidence Collar, which is routinely used for nuclear safeguards on fresh fuel assemblies, that the presence of significant amounts of lithium-6 (10 g) would cause a clearly detectable drop in the coincidence count rate and that the presence of lithium-6 results in a different response than gadolinium which might be used to cover up lithium-6 breeding targets.

If an anomaly was detected, further inspection to identify the reason would be triggered. A screening of 14 plausible nuclear reactions involving interrogating neutrons or gammas revealed that at least one measurement principle for nondestructive identification of lithium-6 in sealed fuel rods, namely nuclear resonance absorption (NRA) with incident gamma rays, appears to be feasible.

Verification of nonremoval is achieved with accountancy complemented by containment and surveillance at tritium-handling facilities. The diversion possibilities to extract inadvertently produced tritium or to remove tritium from existing stocks would make it necessary to put up to 50 facilities worldwide under inspection in addition to those which are already inspected to verify fissile material inventories. Tritium accountancy is already established at these facilities for radiation protection purposes. There is broad experience, and no new technologies are required. Although there are specific technical problems because of tritium being a gas, it is shown in this study that an accountancy capability can be achieved which compares well with the capability of IAEA nuclear safeguards. The precedent for this type is due to the above-mentioned agreement between EURATOM and Canada.

Most facilities which are affected by tritium controls are already under international safeguards for nuclear materials, and most tritium control goals are already covered by routine nuclear safeguards procedures, since tritium can, like plutonium, be produced only in nuclear reactors or in other strong neutron sources. The number of safeguarded facilities might increase by 5%, and only a few measures are necessary in addition to current nuclear safeguards.

As a **conclusion**, this study shows that the verification of international tritium control agreements both for horizontal nonproliferation of tritium and for reversing vertical proliferation (nuclear disarmament) would be technically feasible with an efficiency that is politically acceptable, even if the significant quantity of tritium is defined to be as low as 1 gram. Tritium control could be implemented without adding excessive inspection activities. Existing structures and routine inspection procedures within the nuclear nonproliferation regime can be used.

4.12 Endnotes

1. This is a very conservative assumption, since such a low significant quantity would be even lower than the typical amount of tritium used in one single nuclear warhead, contrary to the definition of significant quantities of fissile materials. A significant quantity of 10 g is more realistic.

Acknowledgments

The work on which this book is based was basically carried out from 1989 to 1998 within IANUS (Interdisziplinäre Arbeitsgruppe Naturwissenschaft, Technik und Sicherheit) at Darmstadt University of Technology. In the following years it was updated for publication. In the first five years, it was funded by the state of Hessia through ZIT (Zentrum für Interdisziplinäre Technikforschung) at Darmstadt University of Technology. In 1995 and 1996 the continuation of the tritium project was made possible by a grant of the Volkswagen Foundation. The same foundation as well as the German Academic Exchange Service provided substantial travel funds during the first years of the project. I owe special thanks to IANUS, which continued to cover various expenses of the project and made its successful conclusion possible.

Harald Müller from the Peace Research Institute Frankfurt (PRIF) provided initial motivation for this project. The work was embedded within a critical mass of physicists at IANUS, namely Egbert Kankeleit, Alexander Glaser, Wolfgang Liebert, Christoph Pistner, Uwe Reichert (until 1990), Annette Schaper (until 1992), and Jürgen Scheffran. They were valuable partners for enlightening discussions about the tritium topic and for fruitful cooperation in the area of nonproliferation and disarmament of nuclear weapons.

I am very grateful to Egbert Kankeleit and his working group at the Institute for Nuclear Physics at Darmstadt University of Technology, who provided a very supportive working environment for this kind of nonestablished interdisciplinary peace research in connection with original research in nuclear physics. I am indebted to Peter Schwalbach, Gisela Buggisch, and Wolfgang Stahn for assisting me in setting up the (n,γ) experiment. Achim Richter, Peter von Neumann-Cosel, and Norbert Huxel supported me while evaluating spectra of nuclear resonance fluorescence that were taken at the S-DALINAC at the Institut für Kernphysik. In 1993, I worked as a guest at the Safeguards Assay Group (N-1) of Los Alamos National Laboratory (LANL) for 14 weeks to do Monte Carlo calculations. I owe special thanks to Egbert Kankeleit for his strong support as supervisor to bring my Ph.D. thesis in nuclear physics to a success.

As part of the interdisciplinary project, the political scientist Lars Colschen did a regime-theoretical study on an international tritium control system. Without his cooperation, the political relevance of the technical part of the study would probably have been much less significant. His Ph.D. thesis was supervised by Ulrich Albrecht at the Free University Berlin.

Thanks are due to Lars Colschen and Wolfgang Liebert for proofreading some chapters and to Regina Hagen for her tremendous support in language editing and final proofreading of the whole manuscript. Thomas Wolf and Herbert Gohla assisted in technical editing. Finally, I thank Taylor & Francis Publishing, Inc., for their professional work in publishing this book.

Appendix A

World tritium facilities, inventories, and production capabilities

This appendix contains tables which summarize the worldwide civilian tritium facilities with their inventories and production capabilities. Each facility type as introduced in Section 2.3 is covered by one table. All numbers are presented by country. These tables are introduced and discussed in Section 2.6.

Military tritium production facilities (facility type 1e) are not included here. They are discussed in Section 2.7 and summarized in Table 2.6.

A.1 Nuclear reactors and special neutron sources (facility type 1)

A.1.1 Nuclear power reactors (facility type 1a,b)

Table A.1 Quantity of tritium which is produced inadvertently by ternary fission and which could be produced by breeding from lithium-6 targets without affecting normal operation in declared civilian power reactors (facility type 1a,b).

country	in operation [a]	under construction [b]	under safeguards [c]	tritium production in operating plants	
	no. of units and total net MWe (in brackets)			inadvertently [d] [10^3 TBq/year]	deliberately [e] *(potential)* [10^3 TBq/year]
Argentina	2 (935)	1 (692)	2	0.56	*9.4–70*
Belgium	7 (5484)	0	all	2.41	*55 –410*
Brazil	1 (626)	1 (1245)	1	0.28	*6.3–47*
Bulgaria	6 (3538)	0	all	1.56	*35– 270*
Canada	21 (14874)	1 (881)	all	8.92	*150–1100*
China	1 (288)	2 (1812)	no	0.13	*2.9–22*
Cuba	0	2 (816)	all	0	*0*
Czech Rep.	4 (1632)	2 (1784)	all	0.72	*16– 120*
Finland	4 (2310)	0	all	1.02	*23– 170*
France	56 (57688)	5 (7125)	no	25.38	*580–4300*
Germany	21 (22559)	6 (3319)	all	9.93	*225–1700*
Hungary	4 (1729)	0	all	0.76	*17– 130*
India	9 (1593)	5 (1010)	2(470)	0.91	*16– 120*
Iran	0	2 (2392)	all	0	*0*
Japan	44 (34238)	9 (8129)	all	15.06	*340–2600*
Kazakhstan	1 (135)	0	no	0.06	*1.4–10*
Korea, Rep. of	9 (7220)	3 (2550)	no	3.18	*72–550*
Lithuania	2 (2760)	1 (1380)	all	1.21	*28–210*

Table A.1 *(continued)*.

country	in operation [a]	under construction [b]	under safeguards [c]	tritium production in operating plants	
	no. of units and total net MWe (in brackets)			inadvertently [d] [10^3 TBq/year]	deliberately [e] *(potential)* [10^3 TBq/year]
Mexico	1 (654)	1 (654)	all	0.29	*6.5–49*
Netherlands	2 (504)	0	all	0.22	*5.1–38*
Pakistan	1 (125)	0	1	0.08	*1.3–9.4*
Romania	0	5 (3155)	all	0	*0*
Russia	28 (18893)	18 (14175)	volunt.	8.31	*190–1400*
Slovak Rep.	4 (1632)	4 (1552)	no	0.72	*16–120*
Slovenia	1 (632)	0	all	0.28	*6.3–47*
South Africa	2 (1842)	0	all	0.81	*18–140*
Spain	9 (7101)	0	all	3.12	*71–350*
Sweden	12 (10002)	0	all	4.40	*100–750*
Switzerland	5 (2952)	0	all	1.30	*30–220*
Taiwan	6 (4890)	0	all	2.15	*49–370*
Ukraine	15 (13020)	6 (5700)	no	5.72	*130–980*
U.K.	37 (12066)	1 (1188)	volunt.	5.31	*120–900*
U.S.	109 (98729)	3 (3480)	volunt.	43.44	*990–7400*
Total	424 (330651)	72 (59720)	186	148	*3,300–25,000*
				0.41 kg/y	*(9.2 -70)kg/y*

[a] The table presents the situation as of 31 December 1992 (IAEA, 1993b).

[b] See IAEA (1993b).

[c] In this column, the situation as of 31 October 1992 is taken into account (IAEA, 1993b).

[d] Inadvertent tritium production via the fuel rod path results from ternary fission and traces of ^{6}Li in the fuel. The production factor for

LWRs is (0.56 to 0.74)$\times 10^3$ TBq/(GW$_e$y) at a 100% capacity factor (IAEA, 1981). The average capacity factor is 68%. Therefore, a production factor of 0.44 TBq/(MW$_e$y) can be reasonably assumed. For HWRs, the maximum production, i.e., at a 100% capacity factor, amounts to 0.814$\times 10^3$ TBq/(GW$_e$y) (IAEA, 1981), whereas the average capacity factor is 74%. Therefore, a production factor of 0.60 TBq/(MW$_e$y) is used here. For the share of HWRs of a country's total nuclear energy capacity, see Table A.2. Other reactor types are in average described best with the production factor of LWRs.

[e] Figures are given for the lithium-6 path assuming that normal reactor operation is not affected. The conversion factor is (10 to 75)$\times 10^3$ TBq/(GW$_e$y) at a 100% capacity factor. Italics indicate the production potential.

Table A.2 Quantity of tritium produced in heavy water power reactors due to neutron capture of deuterium (heavy water path, facility type 1b).

country	in operation[a]	under construction[b]	under safeguards[c]	tritium production in operating plants[d]
	no. of units and total net MWe (in brackets)			[10^3 TBq/year]
Argentina	2 (935)	1 (692)	2	57–84
Canada	21 (14874)	1 (881)	all	900–1340
India	7 (1293)	5 (1010)	2	78–116
Korea, Rep. of	1 (629)	1 (650)	all	39–57
Pakistan	1 (125)	0	1	7.5–11
Romania	0	5 (3155)	all	0
total	32 (17856)	13 (6388)	27	1080–1600
				(3.0–4.4) kg/y

[a] The table presents the situation as of 31 December 1992 (IAEA, 1993b).

[b] (IAEA, 1993b).

[c] (IAEA, 1993a).

[d] The production factor is (60–90)$\times 10^3$ TBq/(GW_ey) at a 100% capacity factor.

A.1.2 Nuclear research reactors (facility type 1c,d)

Table A.3 Quantity of tritium which is produced inadvertently by ternary fission and which could be produced by breeding with lithium-6 targets in research reactors (facility type 1c,d)

country	in operation[a]	under construction[b]	under safeguards[c]	tritium production in operating plants	
	no. of units and total net MW_{th} (in brackets)			inadvertently[d] [TBq/year]	deliberately[e] *(potential)* [10^3 TBq/year]
Algeria	2 (16)	0	2	1.6–3.2	*16–32*
Argentina	5 (3.3)	0	6	0.33–0.66	*3.3–6.6*
Australia	2 (10.1)	0	3	1–2	*10–20*
Austria	3 (10.7)	0	3	1–2	*10–20*
Bangladesh	1 (3)	0	1	0.3–0.6	*3–6*
Belgium	5 (104)	0	5	10–21	*100–210*
Brazil	4 (2.1)	0	3	0.2–0.4	*2–4*
Bulgaria	1 (2)	0	1	0.2–0.4	*2–4*
Canada	11 (142)	1 (0.02)	11	14–28	*140–280*
Chile	2 (15)	0	2	1.5–3	*15–30*
China	13 (161)	0	1	16–32	*160–320*
Colombia	1 (0.03)	0	1	0.003–0.006	*0.03–0.06*
Czech Rep.	4 (10)	1 (0.1)	4	1–2	*10–20*
Denmark	2 (10)	0	2	1–2	*10–20*
Egypt	1 (2)	0	1	0.2–0.4	*2–4*
European Com.	1 (45)	0	1	4.5–9	*45–90*
Finland	1 (0.25)	0	1	0.025–0.05	*0.25–0.5*
France	19 (1025)	0	0	103–206	*1000–2000*
Germany	21 (59)	0	26	5.9–12	*59–120*
Ghana	0	1 (0.03)	0	0	*0*
Greece	2 (5)	0	2	0.5–1	*5–10*

Table A.3 *(continued).*

country	in operation[a]	under construction[b]	under safeguards[c]	tritium production in operating plants	
	no. of units and total net MW_{th} (in brackets)			inadvertently[d] [TBq/year]	deliberately[e] *(potential)* [10^3 TBq/year]
Hungary	2 (5.1)	0	2	0.5–1	*5–10*
India	5 (181)	0	0	18–36	*180–360*
Indonesia	3 (31.1)	0	3	3.1–6.2	*31–62*
Iran	4 (5)	1 (0.03)	4	0.5–1	*5–10*
Israel	1 (5)	0	1	0.5–1	*5–10*
Italy	5 (1.25)	0	7	0.013–0.026	*0.13–0.26*
Jamaica	1 (0.02)	0	1	0.002–0.004	*0.02–0.04*
Japan	19 (189)	3 (30)	23	18.9–37.8	*189–378*
Korea, Dem. Peop. Rep.	1 (8)	0	2	0.8–1.6	*8–16*
Korea, Rep. of	3 (2.25)	1 (10)	3	0.23–0.45	*23–45*
Latvia	1 (25E-6)	0	0	0.0–0.0	*0–0*
Libya	1 (10)	0	1	1–2	*10–20*
Kazakhstan	3 (80)	0	0	8–16	*80–160*
Malaysia	1 (10)	0	1	1–2	*10–20*
Mexico	4 (1)	0	4	0.1–0.2	*1–2*
Morocco	0	1 (2)	0	0	*0*
Netherlands	2 (2.03)	0	3	0.2–0.4	*2–4*
Norway	2 (27)	0	2	2.7–5.4	*27–54*
Pakistan	2 (9)	0	2	0.9–1.8	*9–18*
Peru	2 (10)	0	2	1–2	*10–20*
Philippines	1 (3)	0	1	0.3–0.6	*3–6*
Poland	3 (40)	0	4	4–8	*40–80*
Portugal	1 (1)	0	1	0.1–0.2	*1–2*

Table A.3 *(continued)*.

country	in operation[a]	under construction[b]	under safeguards[c]	tritium production in operating plants	
	no. of units and total net MW_{th} (in brackets)			inadvertently[d] [TBq/year]	deliberately[e] *(potential)* [10^3 TBq/year]
Romania	1 (14)	0	3	1.4–2.8	*14–28*
Russia	20 (399)	3 (400)	1	40–80	*400–800*
Slovak Rep.	0	1 (0.01)	0	0	*0*
Slovenia	1 (0.25)	0	1	0.025–0.05	*0.25–0.5*
South Africa	1 (20)	0	1	2–4	*20–40*
Sweden	2 (51)	0	2	5–10	*50–100*
Switzerland	4 (10)	0	4	1–2	*10–20*
Syria	0	1 (0.03)	0	0	*0*
Taiwan	6 (3.9)	0	6	0.4–0.8	*4–8*
Thailand	1 (2)	0	1	0.2–0.4	*2–4*
Turkey	2 (5.25)	0	2	0.53–1	*5–10*
Ukraine	1 (10)	0	0	1–2	*10–20*
U.K.	9 (0.7)	0	0	0.07–0.14	*0.7–1.4*
U.S.	82 (538)	1 (0.25)	0	54–108	*540–1080*
Uzbekistan	1 (10)	0	0	1–2	*10–20*
Venezuela	1 (3)	0	1	0.3–0.6	*3–6*
Viet Nam	1 (0.5)	0	1	0.05–0.1	*0.5–1*
Yugoslavia	2 (6.5)	0	2	0.65–1.3	*6.5–13*
Zaire	1 (1)	0	1	0.1–0.2	*1–2*
total	304 (3321)	15 (442)	169	330–660	*3300–6600*
				(0.92–1.8) g/y	*(9.2–18) kg/y*

[a] The table presents the situation as of November 1993 (IAEA, 1993b).

[b] See IAEA (1993b).

[c] In this column, the situation as of 31 December 1992 is taken into account (IAEA, 1993a). For some countries this number exceeds the number of operating research reactors because safeguards are not terminated for shutdown reactors.

[d] Inadvertent tritium production via the fuel rod path results from ternary fission and traces of lithium-6 in the fuel. The production factor depends on the reactor type and is (0.19–0.37) TBq/(MW_{th} y) at a 100% capacity factor (IAEA, 1981). At a capacity factor of about 50%, the production of tritium can be estimated at (0.1–0.2) TBq/(MW_{th} y).

[e] Figures are given for the lithium-6 path assuming that the reactor is dedicated to tritium production. The conversion factor is (1.0 to 2.0) $\times 10^3$ TBq/(MW_{th} y) at a 100% capacity factor. Italics indicate the production potential.

Table A.4 Quantity of tritium which is produced inadvertently by capture of neutrons in deuterium in heavy water research reactors (heavy water path, facility type 1d).

country	in operation[a]	under constr.[b]	under safeguards[c]	tritium production in operating plants[d]
	no. of units and total net MW_{th} in brackets			[10^3 TBq/year]
Australia	1 (10)	0	1	0.035–0.15
Canada	2 (177)	0	2	0.9–2.7
China	1 (15)	0	1	0.05–0.22
Denmark	1 (10)	0	1	0.035–0.15
France	1 (57)	0	0	0.20–0.86
Germany	1 (23)	0	1	0.08–0.35
India	2 (140)	0	0	0.5–2.1
Japan	2 (10)	0	2	0.035–0.15
Norway	2 (27)	0	2	0.09–0.41
Russia	1 (0)	0	0	0
U.S.	3 (85)	0	0	0.30–1.3
Yugoslavia	2 (6.5)	0	2	0.023–0.10
total	19 (560.5)	0	12	2.0–8.4 (5.6–23) g/y

[a] The table presents the situation as of December 1992 (Varley *et al.*, 1993).

[b] (Varley *et al.*, 1993).

[c] (IAEA, 1993a).

[d] The production factor is (7–30)$\times 10^3$ TBq/(GW_{th} y) at a 100% capacity factor. Here, 50% are assumed.

A.1.3 Special neutron sources (facility type 1f)

Several techniques are available to generate pulsed neutron beams (Gsponer *et al.*, 1983):

Photoneutron sources: Electron accelerators with a heavy target can generate photoneutron pulses with a maximum instantaneous neutron flux of 1×10^{19} n/cm^2/s (ORELA at Oak Ridge National Laboratory, TN, U.S.).

Electron linac pulsed fast assemblies: The GGA facility (U.S., 1955) generates maximum fluxes of 4×10^{13} n/cm^2/s, IBR-II (U.S.S.R., 1977) generates 1×10^{17} n/cm^2/s.

Spallation neutron sources: Proton beams and heavy targets are used to generate peak fluxes of up to 2×10^{21} n/cm^2/s. See Table A.5. Most data are drawn from Kustom (1981), Carpenter (1979), and Lengeler *et al.* (1993).

Magnetic confinement fusion experiments: A tokamak can produce a flux of about 5×10^{13} n/cm^2/s. Some of these experiments are included in Table A.12.

Inertial confinement fusion: With laser drivers, peak fluxes of 1×10^{22} n/cm^2/s have been achieved (Shiva-Nova II, U.S.).

Dense plasma focus: With a mixture of deuterium and tritium, a maximum flux of 1×10^{20} n/cm^2/s can be generated.

Table A.5 Spallation neutron sources (facility type 1f)

country	name, place (years of operation) * denotes that the facility is under IAEA safeguards.	accelerator and beam data av. beam current [μA]	proton energy [MeV]	pulses per second [Hz]	proton pulse width [μs]	peak theor. neutron flux [$s^{-1}cm^{-2}$]
Canada	TRIUMF	≤100	500			3×10^{15}
Japan	* KENS (Pulsed Neutron Radiation Facility), Tsukuba (since 1980)	10	500	20	0.05	5×10^{14}
U.K.	Harwell (shut down)	9.6	150			
	SNS (Spallation Neutron Source), Rutherford-Appleton Laboratory (since the mid–1980s)	200	800	50	0.45	4.5×10^{15}
U.S.	Nevis (shut down)	1.2	385			2×10^{15}
	ZING-P, Argonne (1974–1977)			10	0.1	5×10^{11}
	ZING-P', Argonne (1977–1980)		200	30	0.1	1×10^{14}
	IPNS-I (Intense Pulsed Neutron Source), Argonne (since early 1980s)	20	500	30–45	0.1	7×10^{14}
	WNR (Weapons Neutron Research), Los Alamos (since 1977)	≤20	800	12–120	3.3	2×10^{13}
	WNR/PSR (Proton Storage Ring), Los Alamos (since 1985)	100	800	12–120	0.27	5×10^{15}
total	5 facilities in operation					

A.2 Fuel fabrication facilities (facility type 2)

A list of commercial and pilot fuel fabrication plants is given in Table A.7. A total of 43 fuel fabrication facilities in 18 countries including some lab-scale facilities are under IAEA safeguards (see Table A.6).

Table A.6 Fuel fabrication plants under IAEA safeguards

country	under safeguards [a]
Argentina	4
Belgium	3
Brazil	1
Canada	5
Denmark	1
Germany	5
India	2
Indonesia	2
Italy	1
Japan	7
Korea, Dem. Peop. Rep.	1
Korea, Rep. of	2
Mexico	1
Romania	1
South Africa	2
Spain	2
Sweden	1
U.S.	2
total	43

[a] This list includes facilities which provide only part of the production process, e.g., pellet manufacturing or final assembly (IAEA, 1993a). Therefore, the number of facilities exceeds the number of fuel fabrication plants given in Table A.7.

Table A.7 Fuel fabrication plants (facility type 2).

country	name of operator; place * indicates safeguarded facility	type of facility[a]	production capacity [t/y][b]	status of planning or operation[c]
Argentina	* Ezeiza	U HWR	300	i.o. since 1982
Belgium	* FBFC; Dessel	U PWR	400	i.o.
	* N.V. Belgonucleaire S. A.; Dessel	MOX LWR	5	i.o. since 1983
	* N.V. Belgonucleaire S. A.; Dessel	MOX LWR	30	extension p.
Brazil	* Nuclei; Resende	U PWR	100	i.o. since 1981
	Nuclei; Resende	U PWR	400	p.
Canada	* Zircatec; Port Hope	U HWR	1200	i.o. since 1956
	* GE Canada, Toronto	U HWR	1050	i.o. since 1967
China	CNEIC; Yibin	U LWR	1	i.o. since 1987
	CNEIC; Yibin	U LWR	150	p. for after 1995
France	FBFC; Pierrelatte	U LWR	500	i.o. since 1984
	FBFC; Romans-sur-Isere	U PWR	750	i.o. since 1979
	SICN; Annecy	U GCR	500	i.o.
	Cogema; Cardarache	MOX FBR/LWR	35	i.o. since 1986
	MELOX, Cogema;	MOX LWR	160	p. for 1995
Germany	* Advanced Nuclear Fuels; Lingen	U LWR	300	i.o. since 1979
India	BARC; Trombay	U HWR	135	i.o. since 1959
	* DAE; Hyderabad	U HWR	250	i.o. since 1971
	* DAE; Hyderabad	U HWR	1500	extension p. for 1995
	DAE; Hyderabad	U LWR	25	i.o.
	NFC; Visakhapatman	U HWR	1500	p. for 1995
Italy	ENEA; Saluggia	U HWR/LWR	60	i.o. since 1986
	CN, Rotondella	U Magnox	200	i.o.

Table A.7 *(continued)*.

country	name of operator; place * indicates safeguarded facility	type of facility[a]	production capacity [t/y][b]	status of planning or operation[c]
	* Fab Nucl, Bosco-Marengo	U BWR	200	i.o. since 1974
Japan	* JFI; Kumatori	U LWR	265	i.o. since 1972
	* Mitsubishi Nuclear Fuels; Tokai-Mura	U PWR	420	i.o. since 1973
	* JFI; Tokai-Mura	U LWR *et al.*	200	i.o. since 1980
	* JNFC; Yokosuka/Uchikawa	U LWR	750	i.o. since 1969
	* PNC; Tokai-Mura (2 facilities)	MOX ATR/FBR	9/5	i.o. since 1972/88
	PNC; Tokai-Mura	MOX LWR/FBR	35	p. for 1995
Korea, Rep. of	* KNFC/KAERI; Taejeon	U PWR	200	i.o. since 1989
Mexico	* JNIN; Centro Nuclear Salazar	U BWR/HWR	2	i.o. since 1992
Pakistan	PAEC; Chashma Barrage	U HWR		i.o. since 1986
Russia	Electrostal, near Moscow	U LWR	700	i.o.
	Electrostal, near Moscow	U RBMK	570	i.o.
	Novosibirsk	U LWR	1000	i.o.
	Paket at Mayak, Ozersk	MOX FBR	1	i.o.
	Paket at Mayak, Ozersk	MOX FBR	60	p.
	facility at RIAR, Dimitrovgrad	MOX FBR	1	i.o.
South Africa	* Pelindaba, Transvaal	U LWR		i.o. since 1981
Spain	* ENU.S.; Juzbado	U LWR	200	i.o. since 1985
	* ENU.S.; Juzbado	U LWR	500	p. for after 1990
Sweden	ABB-Atom; Västeras	U LWR	400	i.o. since 1971
U.K.	BNFL; Springfields	U Magnox	1300	i.o. since 1960
	BNFL; Springfields	U PWR	190	p. for 1995

Table A.7 *(continued)*.

country	name of operator; place * indicates safeguarded facility	type of facility[a]	production capacity [t/y][b]	status of planning or operation[c]
	BNFL; Springfields	U AGR	230	p. for 1995
	BNFL; Springfields	U AGR	263	i.o.
	BNFL; Sellafield	MOX LWR	8	p. for 1992
	BNFL; Sellafield	MOX LWR	100	p. for 1998
	AEA Technology; Dounreay	U RR	500 FE/y	i.o. since 1959
	AEA Technology; Dounreay	U RR	1000 FE/y	p. for 1993
U.S.	Westinghouse Electric; Columbia, SC	U PWR	1250	i.o. since 1969
	* Babcock and Wilcox; Lynchburg, VA	U PWR	400	i.o. since 1972
	Advanced Nuclear Fuels; Richland, WA	U LWR	700	i.o.
	General Electric Co.; Wilmington, NC	U BWR	1100	i.o. since 1982
	Combustion Eng., Hematite, Windsor, CT	U PWR	300	i.o. since 1982
	total		> 16,220	i.o. 43 plants * 22 plants p. 9 plants

[a] U: uranium fuel; MOX: mixed oxide fuel; AGR: advanced gas cooled reactor; BWR: boiling water reactor; FBR: fast breeder reactor; GCR: gas cooled reactor; HWR: heavy water reactor; LWR: light water reactor; PWR: pressurized water reactor; RR: research reactor.

[b] Most figures are taken from IAEA (1988), Nuclear Energy Agency, OECD (1987) and Varley *et al.* (1993). Capacity is given in tonnes of heavy metal. The capacity is not equivalent to the actual production output since the world production capacity considerably exceeds demand. For example, in Europe the average capacity excess is estimated at 40% (Müller and Hossner, 1989). FE stands for fuel elements.

[c] i.o.: in operation; p.: planned

A.3 Separate spent fuel storage facilities (facility type 3)

Table A.8 Separate spent fuel storages (facility type 3).

country	storage capacity [a] [t heavy metal]	number of facilities [b] * indicates safeguarded facility
Argentina	365	1
Belgium	700	(shut down)
Bulgaria	600	1
Canada	13,000	9 * (partly planned)
China	500	(planned)
Czech Rep.	600	(planned)
Finland	1270	1 *
France	15,000	1 *
	?	3
Germany	5900	5 *
India	523	1
Italy	1400	1 *
Japan	5600	1 * (partly planned)
Korea, Rep. of	3000	(planned)
Russia	13,000	2
Slovakia	600	1
South Africa	100	1 *
Spain	5500	(planned)
Sweden	5000	1 * (partly planned)
Switzerland	700	1 *
U.K.	11,700	1 *
Ukraine	1900	1
U.S.	17,600	3 (partly planned)
total	>100,000	38 facilities 23 safeguarded

[a] The numbers are taken from IAEA (1993d), Varley *et al.* (1993), or as given by the U.S. Department of Energy for 1988. No spent fuel was believed to be stored at that time in Algeria, Australia, Austria, Chile, China, Columbia, Cuba, Denmark, Egypt, Greece, Indonesia, Iran, Iraq, Israel, Jamaica, Dem. Peop. Rep. of Korea, Libya, Malaysia, Mexico, Norway, Peru, Philippines, Poland, Portugal, Romania, Thailand, Turkey, Venezuela, Zaire. A question mark (?) denotes that no data are available.

[b] See also (IAEA, 1993a).

A.4 Reprocessing plants (facility type 4)

Table A.9 Reprocessing plants (facility type 4).

country (name, place), time of operation [a]	fuel type [b]	reported and projected capacity [t/y] [c]		cumulative fuel reprocessing through 1991 [t] [d]	mean burn-up [$GW_{th}d/t$]	tritium released by end of 1991 [g] [e]	throughput of tritium [g/y] in 2000 [f]
		1991	2000				
Argentina (Ezeiza), since 1989	ox LWR	5	5	<10	25 (?)	<0.6 (?)	0.64–0.86
Belgium (Eurochemic, Mol), 1966–74	ox GCR	0	0	10	0.9–1.5	0.02–0.032	0
	ox HWR	0	0	70	4–6	0.77–1.2	0
	ox PWR	0	0	72	12.9–21	2.0–3.2	0
	ox BWR	0	0	30	6–17.3	0.41–1.2	0
Brazil (IPEN, Sao Paulo), since 1983	ox LWR	lab-scale	lab-scale				
France (UP-1, Marcoule), since 1958 (*)	met GCR	400	0	5200	≃3	33	0
France (UP-2, La Hague), 1966–87 (*)	met GCR	0	0	4895	≃3	31	0
France (UP-2 + Oxid-Head-End, La Hague), since 1976 (*)	ox LWR MOX LWR/FBR	400	800	3671	20	157	51–69
France (UP3, La Hague), due 1992 (*)	ox LWR	800	800	570	25	30.5	51–69
France (Marcoule)	FBR	5	5	≃20	10 (?)	0.62 (?)	0.15
Germany (WAK, Karlsruhe), 1971-90 *	ox L/HWR	(35)	0	208	20	8.9	0

Table A.9 *(continued)*.

country (name, place), time of operation [a]	fuel type [b]	present and projected capacity [t/y] [c] 1991	2000	cumulative fuel reprocessing through 1991 [t] [d]	mean burn-up [$GW_{th}d/t$]	tritium released by end of 1991 [g] [e]	throughput of tritium [g/y] in 2000 [f]
India (BARC, Bombay), since 1964, expanded 1985	ox U	50	60				3.9–5.1
India (PREFRE, Tarapur), since 1977 *	ox CANDU	100	100	≃140	≃6	2.3	1.64
India (Kalpakkam), since 1986	ox CANDU	1125	1125				18.5
Italy (ENEA, Saluggia), since 1969 *	ox	10	10				0.64–0.86
Italy (Rotondella), since 1969 *	ox	4	4				0.28–0.34
Japan (Tokai-Mura), since 1981 *	ox LWR	90	210	638	19.5	27	13.5–18
Japan (Rokkasho-Mura), due 2000	ox LWR	0	800	0	0	0	51–69
Dem. P. Rep. Korea (Yongbyon) *	ox	lab-scale					
Pakistan (Pinstech, Rawalpindi), since 1960		lab-scale					
Pakistan (Pinstech, Rawalpindi), since 1981		pilot pl.					
Russia (Mayak, Ozersk), since 1978	ox LWR	600	600	≃2800	≃25	150	39–51

Table A.9 *(continued)*.

country (name, place), time of operation [a]	fuel type [b]	present and projected capacity [t/y] [c]		cumulative fuel reprocessing through 1991 [t] [d]	mean burn-up [$GW_{th}d/t$]	tritium released by end of 1991 [g] [e]	throughput of tritium [g/y] in 2000 [f]
		1991	2000				
Russia (RT-1, Chelyabinsk), since 1971	ox LWR	400			30		26–34
Russia (RT-2, Krasnoyarsk), planned since 1975	ox LWR	–	(1100)	–	–	–	(180–230)
U.K. (B205, Sellafield), since 1964 (*)	met GCR	1500	1500	21350	≃3	137	96–129
U.K. (B204/205, Sellafield), 1969-1973 (*)	ox LWR	(300)	0	73	12	1.9	0
U.K. (THORP, Sellafield), since 1995 (*)	ox LWR	0	850	0	0	0	55–73
U.K. (DNPDE, Dounreay), since 1959 (*)	ox MTR	<1					<0.12
U.K. (DNPDE, Dounreay), since 1980	MOX FBR	7	7	24	10 (?)	0.74 (?)	0.15
U.S. (NFS, West Valley), 1966-1971	ox L/HWR	(300)	0	624	8.5	11.4	0
total (20 operating plants, 2 under construction)		≃5400	≃8000	≃40,000		600	480–630

[a] met: metal fuel; ox: oxide fuel; BWR: boiling water reactor; FBR: fast breeder reactor; GCR: gas cooled reactor; HWR: heavy water reactor; LWR: light water reactor; PWR: pressurized water reactor. UP2 was enlarged and renamed UP2-800.

[b] * denotes that the facility is under safeguards, (*) indicates partly safeguarded facilities (IAEA, 1993a).

[c] Work is normally performed in campaigns and not in continuous operation. Therefore, the design capacities do not reflect adequately the actually reprocessed quantities. See Varley *et al.* (1993).

[d] Figures are partly taken from Berkhout (1992), Müller and Hossner (1991) and previous editions. Figures given in literature are extrapolated to the end of the year 1991.

About 300,000 to 800,000 t U with low burn-up (about 600 MWd/tU) had been reprocessed for weapons production in military facilities worldwide.

[e] The following production rates are assumed in order to calculate the amount of tritium released (IAEA, 1981). BWR and PWR: 1.8 g/(GWey); FBR: 2.6 g/(GWey); AGR: 1.75 g/(GWey); HWR: 2.3 g/(GWey). The figures are not decay-corrected.

Not all of the tritium is released to the environment, however. For comparison the normalized emission rates (Marcoule (met GCR): 350 TBq/(GW_ey), La Hague (UP2-400): 290 TBq/(GW_ey), and Sellafield (met GCR): 700 TBq/(GW_ey). See United Nations Scientific Committee on the Effects of Atomic Radiation (UNSCEAR), 1988) are used to estimate total tritium releases: 42 g, 160 g, and 340 g, respectively.

Since the burn-up is low for the production of nuclear weapons plutonium, the amount of tritium released by these military reprocessing operations is between 1 and 2.5 kg tritium worldwide.

[f] Higher burn-up is assumed for the fuel reprocessed in 2000 (30 to 40 GW_{th}d/t U for LWRs). In reality, tritium throughput will be lower than stated in this column since the nominal reprocessing capacity will not be fully utilized and tritium decay is not accounted for.

For comparison, the normalized emission rates as given in the footnote above (United Nations Scientific Committee on the Effects of Atomic Radiation (UNSCEAR), 1988) are used to estimate tritium releases from La Hague (UP2-800): 53-71 g, and Sellafield (met GCR): 144 g.

A.5 Final disposal sites for nuclear waste (facility type 5)

Table A.10 Nuclear waste disposal sites (facility type 5).

country	name of operator; place [a]	type of facility [b]	waste form [c]	storage capacity [m^3] [d]	status of planning or operation [e]
Argentina	Siera del Medio	UF	HLW		s.i.
Australia		SLBF	LLW		p.
Belgium	ONDRAS/NIRAS; (no site selected)	SLBF	A		s.i. stopped
	ONDRAS/NIRAS; (no site selected)	DUF	B, C		p. 2020
	ONDRAS/NIRAS, HADES; Mol-Dessel	DUF	B, C		i.o. since 1983
Brazil		SLBF	LLW		p.
Bulgaria		ESF	LLW		i.o.
		SLBF	LLW		i.o.
Canada	AECL; Chalk River Nuclear Laboratories	SLBF	LLW, ILW		u.c. since 1989
	AECL; (no site selected)	DUF	HLW	63,000	u.l.p.; p. 2015
Chile		ESF	LLW		i.o.
China		SLBF	LLW		p.
	Gobi	DUF	HLW		p.
Cuba		ESF	LLW		i.o.
		SLBF	LLW		p.
Egypt		ESF	LLW		p.
Finland	TVO; Olkiluoto	DUF	LLW, ILW		p. for 1992
	IVO, Loviisa	DUF	LLW, ILW		p. for 2000
	TVO; (no site selected)	DUF	HLW		p. for 2020
France	ANDRA; Centre de la Manche	SLBF	S-LW	485,000	i.o. 1969 - 91
	ANDRA; Soulaines-Dhuys, Aube	SLBF	S-LW	1,000,000	i.o. since 1991
	ANDRA; (no site selected)	DUF	L-LW	80,000	s.s.u.w. since 1987

Table A.10 *(continued)*.

country	name of operator; place [a]	type of facility [b]	waste form [c]	storage capacity [m^3] [d]	status of planning or operation [e]
Germany	DBE; Asse	DUF	LLW, ILW	125,000 drums	i.o. 1967 - 78
	DBE, Schacht Konrad; Salzgitter	UF	LHGW	650,000	u.l.p., p. for 1996
	DBE; Gorleben	DUF	HHGW		p. for 2008
	Morsleben	DUF	LLW, ILW	1,000,000	i.o. 1981 - 91
Hungary	northeast of Budapest	SLBF	LLW, ILW		i.o.
Indonesia		ESF	LLW		p.
		SLBF	LLW		p.
		DUF	HLW		p.
Italy	ENEA	SLBF	LLW		
	ENEA	DUF	HLW		
Japan	J. Nuclear Fuel Industries; Rokkasho-Mura	SLBF	LLW	200,000	u.c.; p. for 1992
	extension			400,000	later
	J. Nuclear Fuel Service; (no site selected)	DUF	HLW		s.s.u.w. since 1985
Jordan		SLBF	LLW		p.
Korea, Rep. of	(no site selected)	DUF	LLW, ILW, HLW		s.s.u.w.
Malaysia		ESF	LLW		p.
Mexico		SLBF	LLW, (ILW)		i.o.
		SLBF	LLW, (ILW)		p.
Netherlands	COVRA	DUF	LLW, ILW, HLW		s.s.u.w.
Norway		SLBF	LLW		i.o.
Pakistan		SLBF	LLW		i.o.
Poland		ESF	LLW		i.o.
		DUF	LLW		p.

Table A.10 *(continued)*.

country	name of operator; place [a]	type of facility [b]	waste form [c]	storage capacity [m^3] [d]	status of planning or operation [e]
Russia	several (near reactor sites)	ESF	LLW		several i.o.
	(site selected in granite)	DUF	HLW		p.
	(no site selected in other materials)	DUF	HLW		s.s.u.w.
South Africa	Vaalputs	SLBF	LLW, ILW		i.o. since 1988
	(no site selected)	DUF	HLW		p.
Spain	ENRESA; El Cabril, Sierra Albarrana	UF	LLW, ILW	58,000	p. 1990
	ENRESA, (no site selected)	DUF	HLW		s.s.u.w.
Sweden	Oskarsham	SLBF	LLW	9,000	i.o. since 1986
	SKB, SFR; Forsmark	SSBF	LLW (ILW)	60,000	i.o. since 1988
	extension			30,000	later
	SKB, SFL; (no site selected)	DUF	HLW	2,600	p. 2020
Switzerland	NAGRA; Wellenberg	UF	LLW, ILW	100,000	s.i., p. 2000
	NAGRA; (no site selected)	DUF	HLW		p. 2020
U.K.	BNFL; Drigg, Cumbria	SLBF	LLW	600,000	i.o. since 1959
	U.K. NIREX Ltd; Sellafield	DUF	LLW, ILW	800,000	p. for 2005
U.S.	DOE, WIPP; Carlsbad, New Mexico	DUF	HLW	178,000	u.c., test phase
	Yucca Mountain, Nevada	DUF	HLW		s.i., p. 2010
	six sites[f]	SLBF	LLW		3 i.o.; 3 closed
Zambia		SLBF	LLW		i.o.
		ESF	LLW		p.

[a] AECL: Atomic Energy of Canada Limited; ANDRA: Agence Nationale pour la gestion des Dechets Radioactifs, Paris; BNFL: British Nuclear Fuel Limited, Risley, Warrington; COVRA: Centrale Organisatie Voor Radioactief Afval, Petten; DBE: Deutsche Gesellschaft zum Bau und Betrieb von Endlagern für Abfallstoffe mbH, Peine; ENEA: Comitato Nazionale per la Ricerca e per lo Sviluppo dell'Energia Nucleare e delle Energie Alternative; ENRESA: Empresa National de Residuos Radioactivos, Madrid; HADES: High Activity Disposal Experimental Site; IVO: Imatran Voima Oy; NAGRA: Nationale Genossenschaft für die Lagerung radioaktiver Abfälle, Baden;

NIREX: Nuclear Industry Radioactive Waste Executive, Harwell; ONDRAF: Organisme National des Dechets Radioactifs et des Matieres Fissiles, Brussels; OPLA: a research guiding committee; SFL: Final disposal for HLW; SKR: Final disposal for LLW/ILW; SKB: Swedish Nuclear Fuel and Waste Management Company; TVO: Teollisuuden Voima Oy; WIPP: Waste Isolation Pilot Plant (for military α-waste).

[b] ESF: engineered surface facility, SLBF: shallow land burial facility, SSBF: sub-sea-bed facility, UF: underground facility, DUF: deep underground facilities.

[c] LLW: low level solid waste, ILW: intermediate level waste, HLW: high level waste, A, B, C: categories defined by Belgium, S-LW: short-lived waste, L-LW: long-lived waste, LHGW: low heat generating waste, HHGW: highly heat generating waste.

[d] For HLW a density of $\rho = 3$ g/cm^3 can be assumed to convert storage capacity figures officially given in tonnes. This factor was applied for the facilities of Canada and Sweden.

[e] p.: planned for (start of operation in the year) ..., s.s.u.w.: site selection under way, s.i.: site investigation, u.l.p.: undergoing license procedure, u.c.: under construction, u.t.o.: under test operation, i.o.: in operation.

[f] In the U.S., each state is responsible for managing and disposing of its own low-level radioactive waste. Still in operation are three of six disposal facilities, namely at Barnell (SC), Richland (WA), and Beatty (NV). Closed are Maxey Flats (KY), West Valley (NY), and Sheffield (IL).

A.6 Detritiation facilities (facility type 6)

Table A.11 Detritiation facilities for tritiated (heavy) water (facility type 6)

country	name, place (operating since)	feed rate capacity [l/h] [a]	maximum activity [TBq/l] [b]	tritium production capacity [g/year]
Belgium	prototype cell, CEN/SCK Mol (1988, tests)	0.004	?	<1
Canada	Chalk River Nuclear Laboratory (1988)	20	1	140[c]
	Tritium Recovery Facility (TRF), Darlington (1988)	370	1.1 (initially) 0.1 (in equilibrium)	2500[d]
France	Grenoble[e] (1972)	20	?	16
India	Bhabha Atomic Research Centre (pilot plant)	?	?	?
U.S.	Tritium Aqueous Waste Recovery System, Mound (1986)	1.2	0.002	0.06
	total			<2700

[a] CFFTP (1988).

[b] CFFTP (1988).

[c] Dinner (1988). The theoretical tritium production capacity of 4 kg/year cannot be reached, because the necessary feed is not available.

[d] This figure assumes uninterrupted operation. More realistic would be an average of 2200 to 2300 g/y with about 3500 g/y within the first few years of operation when more highly tritiated water is available and processed first. The estimate was made in 1988 on the basis of Ontario Hydro's nuclear capacity at that time of 12 GW_e (with an additional 3.5 GW_e under construction), assuming a capacity factor of 80% (CFFTP, 1988). From 1988 to June 1993, only 5.7 kg were extracted at Darlington, according to Fusion Canada—Bulletin of the National Fusion Program, Issue 21, August 1993. This low extraction figure is mainly due to a long shutdown caused by various operating problems shortly after plant start-up.

[e] Pautrot and Arnauld (1975).

A.7 Tritium research and storage facilities (facility type 7)

Table A.12 Tritium research and storage facilities with inventory or annual throughput > 1 gram (facility type 7).

name	place, country	year of first tritium operation	inventory in 1994 [g]	maximum inventory [g]
ANL Tritium Process Development Group of the Chemical Engineering Division	Argonne National Lab., U.S.			
Bruyeres-le-Chatel Research Center of the French Atomic Energy Commission CEN	France		>1	
Burning Plasma Experiment	Princeton Plasma Phys. Lab., U.S.	planned		
CRNL (Chalk River Nuclear Laboratory)	Chalk River, Canada		2	
ETHEL (European Tritium Handling Experimental Laboratory)	Joint Research Center Ispra, Italy	1994	>1	100
IGNITOR-2 (Ignition Tokamak Reactor Experiment)	Canada	mid-1990s		
INEL Tritium Research Facility (Idaho National Engineering Laboratory)	Idaho, U.S.			
ITER (International Thermonuclear Experimental Reactor)	no decision made	planned		5000
JET (Joint European Torus)	Culham, England	1991	0.24 [a]	90
JT-60 (JAERI-Tokamak-60)	Japan	planned		
KAPL (Knolls Atomic Power Laboratory)	U.S.			
KMS Fusion Inc., tritium handling facility				
LLNL Fusion Engineering Research Facility (NOVA, Tritium Facility and Fusion Target Development Facility)	Lawrence Livermore National Laboratory, U.S.		>1	
Mol Nuclear Research Center	Belgium	planned		
OHTHE-3 (Ohmically Heated Tokamak Experiment)	Canada	mid-1990s		
Ontario Hydro Research Div.	Toronto, Canada		50	
ORNL IRML (Isotope Research Material Laboratory)	Oak Ridge, U.S.	1969	100	

Table A.12 *(continued)*.

name	place, country	year of first tritium operation	inventory in 1994 [g]	maximum inventory [g]
ORNL (Oak Ridge National Laboratory), storage for rejected light sources	Oak Ridge, U.S.	1979	6	
Sandia Tritium Plasma Experiment	Albuquerque, U.S.			
SIN	Switzerland	1987	20	
SRL (Savannah River Laboratory)	U.S.		>1	
T-15	Russia	planned		
TAWRS (Tritium Aqueous Waste Removal System at Mound Laboratory)	Miamisburg, U.S.		>1	
TFTR (Tokamak Fusion Test Reactor)	PPPL, U.S.	1993	$\simeq$1	50
TLK (Tritiumlabor Karlsruhe)	Karlsruhe, Germany	1994	>1	200
TPL (Tritium Process Laboratory) at JAERI (Japan Atomic Energy Research Institute)	Tokai, Ibaraki, Japan	1986	60	
TRL (Tritium Research Laboratory)	Sandia National Lab., U.S.	1977	>1	120
TSF (Tritium Salt Facility)	Los Alamos, U.S.		>50	
TSTA (Tritium Systems Test Assembly)	Los Alamos Nat. Lab., U.S.	1982	110	
University of Toronto Tritium Laboratory for fusion research	Toronto, Canada			
23 facilities in operation		in 1994	> 400	1000
7 facilities planned				> 5000

[a] Two shots with 0.1 g of tritium each were performed on November 9, 1991. About 0.4 g of tritium were on-site during that experiment, because slightly more tritium has to be in the fueling system. For future experiments, the total tritium content of the recycling system will be 90 g during the operation phase (Lässer, 1989).

A.8 Commercial tritium manufacturers and trade companies (facility type 8).

Table A.13 Large commercial tritium manufacturers and trade companies (facility type 8).

country	name, place	main products	throughput [g/year]
Canada	Canadian Fusion Fuel Technology Project (CFFTP)	trade, tritium gas	10–50
China	Second Ministry of Machine Building Industry [a]	powder for watches, tritium lamps	
France	CEA/VALDUC	tritiated targets	
France	Tritium labeling facility, CEN Saclay [b]	labeled molecules	1
Germany	Hans Gutekunst Leuchtstoffe, Villingen-Schwenningen	GTLS [c] and luminous compounds	
Russia	Radium Institute, St. Petersburg	luminescent signs	
Russia	Alkor Technologies, St. Petersburg	trade, tritium gas	
Switzerland	Radium Chemie AG, Teufen/AR		$\simeq$5
Switzerland	Mb-Microtec AG, Niederwangen/BE		$\simeq$40
U.K.	Amersham Buchler		
	Beta Lighting Ltd.		
	Brandhurst Co. Ltd.	self-powered light sources	
	Saunders Roe Development Ltd.	self-powered light sources	
	Surelite Ltd.		
U.S.	American Atomics (closed in 1978)	self-powered light sources	
	Self-Powered Lighting Inc., NY	self-powered light sources	
	Edlow	luminous paint	
	Lawrence Berkeley National Laboratory, National Tritium Labelling Facility	self-powered light sources	
	Oak Ridge National Laboratory sales program	trade, tritium gas	200
	Oak Ridge National Laboratory RL lights program	radioluminescent (RL) lights	5 [d]
	Safety Light	luminous paint	
	SRB Technology		
world	in total 21 large companies in business		> 300

[a] Joint Publications Research Service (1988).

[b] Bull. d'Informations Scientific et Techniques, (1973).

[c] GTLS: gaseous tritium light sources.

[d] The program for the development and fabrication of RL lights at ORNL began in 1979. By 1988, they produced RL lights containing a total of more than 18.5 PBq tritium (Kobisk, 1989).

References

Berkhout, F., *et al.* (1992) Disposition of separated plutonium. *Science and Global Security*, 3, 1–53.

Carpenter, J.M., *et al.* (1979) Pulsed spallation neutron sources. *Physics Today*, 42, December.

CFFTP (1988) *Tritium Supply for Near-Term Fusion Devices.* CFFTP-G-88024, May.

Dinner, P., *et al.* (1988) Tritium Technology Development in EEC Laboratories. Contributions to Design Goals for NET. *Fusion Technology*, 14, 418.

Gsponer, A., Jasani, B. and Sahin, S. (1983) Emerging nuclear energy systems and nuclear weapon proliferation. *Atomkernenergie/Kerntechnik*, 43, 169–174.

IAEA (1981) Handling of Tritium-Bearing Wastes. Technical Report Series, 203, Vienna.

IAEA (1988) Nuclear Fuel Cycle Information System. Vienna.

IAEA (1991a) World Survey of Activities in Controlled Fusion Research. Nuclear Fusion Special Supplement, Vienna.

IAEA (1991b) Safe Handling of Tritium. Review of Data and Experience. *Technical Report Series* 324. Vienna.

IAEA (1993a) The Annual Report for 1992. Vienna.

IAEA (1993b) Nuclear Power Reactors in the World. Reference Data Series 2, Vienna.

IAEA (1993c) Nuclear Research Reactors in the World. Reference Data Series 3-d Edition, IAEA, Vienna.

IAEA (1993d) IAEA Yearbook 1993. Vienna.

Joint Publications Research Service (JPRS) (1988) Selections from China Today: Nuclear Industry. Science and Technology, China. Report JPRS-CST-88-002, January, Washington.

Kobisk, E.H., *et al.* (1989) Tritium-Processing Operations at the Oak Ridge National Laboratory with Emphasis on Safe-Handling Practises. *Nucl. Instr. Methods*, A282, 329–340.

Kustom, R.L. (1981) Intense pulsed neutron sources. *IEEE Trans.*, NS-28, 3115–3119.

Lässer, R. (1989) *Tritium and Helium-3 in Metals.* Berlin.

Lengeler, H., Richter, D. and Springer, T. (1993) Spallationsneutronenquelle – Das nächste große europäische Gemeinschaftsprojekt? *Physikalische Blätter*, 49, 1021–1023.

Müller, W.D. and Hossner, R. (eds) (1989) *Jahrbuch für Atomwirtschaft 1989.* Düsseldorf.

Müller, W.D. and Hossner, R. (eds) (1991) *Jahrbuch für Atomwirtschaft 1991.* Düsseldorf.

Nuclear Energy Agency, OECD (1987) Electricity, Nuclear Power and Fuel Cycle in OECD Countries. Main Data. Paris.

Pautrot, P. and Arnauld. (1975) Tritium Extraction Plant of the Laue Langevin Institute. *Trans. Am. Nucl. Soc.*, 20, 202.

Prescott, R.F. (1988) European Tritium Handling Experimental Laboratory (ETHEL). *The Nuclear Engineer*, 29, No. 3, 101.

United Nations Scientific Committee on the Effects of Atomic Radiation (UNSCEAR) (1988) Sources and Effects of Ionizing Radiation. United Nations.

Varley, J., Dingle, A. and Gee, S.C. (1993) *World Nuclear Industry Handbook 1993.* Sutton.

Appendix B

Abbreviations

AECL	Atomic Energy of Canada Limited
AFRS	away from reactor storage
AGR	advanced gas cooled reactor
BNFL	British Nuclear Fuel Limited, Risley, Warrington
BfS	Bundesamt für Strahlenschutz, Salzgitter
BPR	burnable poison rod
BWR	boiling water reactor
CD	Conference on Disarmament
CEA	Commissariat à l'Energie Atomique
CFFTP	Canadian Fusion Fuels Technology Project
CoCom	Coordinate Committee for Multilateral Export Control
COGEMA	Compagnie Generale des Matieres Nucleaires
CRNL	Chalk River Nuclear Laboratory
C/S	containment and surveillance
DOE	U.S. Department of Energy
ETHEL	European Tritium Handling Experimental Laboratory
FBR	fast breeder reactor
GC	gas chromatography
GCR	gas cooled reactor
GTLS	gaseous tritium light source
HEU	highly enriched uranium
HLW	high level radioactive waste
HTGR	high temperature gas cooled reactor
HTR	high temperature reactor
HWR	heavy water reactor
IAEA	International Atomic Energy Agency, Vienna
ICF	inertial confinement fusion
ICO	Integrated Cutoff
ICRP	International Commission on Radiological Protection
ILW	intermediate level radioactive waste
INF	Intermediate Nuclear Forces Treaty
ITCS	International Tritium Control System

JET	Joint European Torus
LLNL	Lawrence Livermore National Laboratory
LLW	low level radioactive waste
LMFBR	liquid metal fast breeder reactor
LWBR	light water breeder reactor
LWR	light water reactor
MBA	Material Balance Area
MTR	Material Testing Reactor
MUF	Material Unaccounted For
NCC	Neutron Coincidence Collar
NCRP	United States National Committee on Radiation Protection
NDA	nondestructive analysis
NEA	Nuclear Energy Agency of the OECD, Paris
NMMG	Nuclear Materials Management Group
NOL	normal operating loss
NPT	Nonproliferation Treaty
NRA	Nuclear Resonance Absorption
NRC	United States Nuclear Regulatory Commission
NRF	Nuclear Resonance Fluorescence
NSG	Nuclear Suppliers Group
NTM	national technical means
OECD	Organization for Economic Cooperation and Development, Paris
ORNL	Oak Ridge National Laboratory
PFNAA	Prompt Fast Neutron Activation Analysis
PIV	Physical Inventory Verification
RPE	rapid power excursion
PVT/MS	pressure volume temperature / mass spectroscopy
PWR	pressurized water reactor
SBR	seed-blanket reactor
SNLL	Sandia National Laboratories Livermore
SNM	special nuclear material
SNS	spallation neutron sources
SQ	significant quantity
SRP	Savannah River Plant
SSAC	State's System of Accounting and Control
START	Strategic Arms Reduction Treaty
STP	standard temperature and pressure
StrlSchV	Strahlenschutzverordnung, the German radiation protection ordinance
TAWRS	Tritium Aqueous Waste Recovery System
TERS	Tritium Effluent Removal Systems
TFTR	Tritium Fusion Test Reactor
TLK	Tritium Laboratory Karlsruhe
TRF	Tritium Removal Facility
TSTA	Tritium Systems Test Assembly
UN	United Nations

UNSCEAR	United Nations Scientific Committee on the Effects of Atomic Radiation
UNSCOM	United Nations Special Committee
USAEC	U.S. Atomic Energy Commission
WNR	Weapons Neutron Research, Los Alamos

Subject index